Sowing Beyond the State

NGOs and Seed Supply in Developing Countries

Elizabeth Cromwell and
Steve Wiggins
with Sondra Wentzel

Overseas Development Institute

A CIP Publication data record may be obtained from the British Library.

ISBN 0-85003-193-1

© Overseas Development Institute 1993

Published by the Overseas Development Institute,
Regent's College, Inner Circle, Regent's Park,
London NW1 4NS

Cover photograph by Godfrey Cromwell

Typeset at the Overseas Development Institute
Printed by the Russell Press Ltd, Nottingham
on recycled paper

Contents

4

Tables, Figures and Diagrams

Acronyms

AA—N	ActionAid–Nepal
AAASA	Association for the Advancement of Agricultural Sciences in Africa
AATG	ActionAid–The Gambia
ABS	Agricultural Bank of Sudan
ACORD	Agency for Cooperation and Research in Development
AIC	Agricultural Inputs Corporation (Nepal)
ARD	Associates in Rural Development (US)
ASC	Agricultural Service Centre (Nepal)
ASS	African Seeds of Survival (Ethiopia)
BARC	Bangladesh Agriculture Research Council
BAU	Bangladesh Agricultural University
CESA	Ecuadorian Centre for Agricultural Services
CIC	Centro Internazionale Crocevia
CIP	International Potato Centre
CGIAR	Consultative Group for International Agricultural Research
CIAT	International Centre for Tropical Agriculture
CIDA	Canadian International Development Authority
CLADES	Latin American Consortium for Agroecology and Development
CSB	Community Seed Bank (Philippines)
CSSTD	Central Seed Science and Technology Division (Nepal)
DAR	Department of Agricultural Research (The Gambia)
DAS	Department of Agricultural Services (The Gambia)
DOA	Department of Agriculture (Nepal)
EEC	European Economic Community
ELC	Environmental Liaison Centre International
EMI	Embu–Meru–Isiolo project (Kenya)
ENDA	Environment and Development Activities–Zimbabwe
ENS	Empresa Nacional de Sementes (Mozambique)
ESC	Ethiopia Seeds Corporation
FAO	UN Food and Agriculture Organisation
F1	First filial generation hybrid
FFHC	Freedom From Hunger Campaign
FV	Farmers' variety
GATT	General Agreement on Tariffs and Trade
GCU	The Gambia Cooperative Union
GDP	Gross Domestic Product
GNP	Gross National Product
GPMB	Gambia Produce Marketing Board
GPSN	Niassa Seed Production Board (Mozambique)
GRAIN	Genetic Resources Action International
GSM	Good Seed Mission (The Gambia)
GTZ	Deutsche Gesellschaft für Technische Zusammenarbeit
HYV	High Yielding Variety
IARC	International Agricultural Research Centre
ICDA	International Coalition for Development Action
IFOAM	International Federation of Organic Agricultural Movements
INIAP	National Agricultural Research Institute (Ecuador)

IRRI	International Rice Research Institute
ISTA	International Seed Testing Association
IVO	Development Research Centre (The Netherlands)
KHAP	Koshi Hills Agriculture Project (Nepal)
KMP	Peasant Movement of The Philippines
KSHP	Kebkabiya Smallholders' Project (Sudan)
LEC	Livelihood Enhancement Committee (The Philippines)
MASIPAG	Farmers–Scientists Partnership for Agricultural Development (The Philippines)
MCC	Mennonite Central Committee
MIND	Mindoro Institute for Development Inc. (The Philippines)
MNC	Multi-National Corporation
MONAP	Mozambique Nordic Agriculture Programme
MV	Modern Variety
NEF	Near East Foundation
NGO	Non-Governmental Organisation
NPK	Nitrogen–Phosphorous–Potassium fertiliser
NRI	Natural Resources Institute (UK)
NSA	National Seed Administration (Sudan)
NSB	National Seed Board (Bangladesh)
ODA	Overseas Development Administration (UK)
ODI	Overseas Development Institute (UK)
OECD	Organisation for Economic Cooperation and Development
OFSP	On-Farm Seed Production Project (The Gambia and Senegal)
OP	Open-pollinated variety
PAC	Pakhribas Agricultural Centre (Nepal)
PAI	Paraná Agronomy Institute (Brazil)
PARE	Partnership in Agricultural Research and Extension (Bangladesh)
PBR	Plant Breeders' Rights
PGR	Plant genetic resources
PGRC/E	Plant Genetic Resources Centre/Ethiopia
PPS	Private Producer-Sellers Project (Nepal)
RAFI	Rural Advancement Foundation International
RENAMO	Mozambique National Resistance
SAM	Sahabat Alam Malaysia
SCF	Save the Children Federation (US)
SEAN	Seed Entrepreneurs Association of Nepal
SEARICE	South East Asia Regional Institute for Community Education
SEMOC	Sementes de Mozambique Lda
SIBAT	Spring of Science and Technology (The Philippines)
SPG	Seed Producer Group (Nepal)
STIP	Seed Technology and Improvement Program (Nepal)
STU	Seed Technology Unit (The Gambia)
UN	United Nations
UPLB	University of The Philippines at Los Banos
USAID	United States Agency for International Development
USC	Unitarian Service Committee (Canada)
VRC	Village Relief Committee (Sudan)
WIPO	World Intellectual Property Organisation
Z–SAN	Zimbabwe Seeds Action Network

Preface and Acknowledgements

This book presents the results of a study of the seed activities of 18 non-governmental organisations and other development agencies in Asia, Africa and Latin America. The study was undertaken by the Overseas Development Institute (ODI) during 1991 and 1992 and funded by the Natural Resources and Environment Department of the UK Overseas Development Administration (ODA).

The study grew out of two ongoing research projects at ODI: Elizabeth Cromwell's work on the structure and economics of the seed sector in developing countries, and John Farrington's work on the role of NGOs in agricultural research and extension in developing countries.

The authors wish to express their sincere thanks to all the staff of the many development agencies who gave up time to provide valuable information and advice during the course of the research. Thanks are also due to the ODA, which funded the cost of the research. At ODI, thanks are due to John Farrington, Research Fellow, who provided valuable advice on the scope and organisation of the study, and to Geraldine Healy who prepared this book for publication.

Elizabeth Cromwell is an Agricultural Economist and a Research Fellow at the Overseas Development Institute. She was research leader for this study. Steve Wiggins is an Agricultural Economist and a lecturer in the Department of Agricultural Economics and Management at the University of Reading. Sondra Wentzel is a Development Anthropologist and is currently an employee of Deutsche Gesellschaft für Technische Zusammenarbeit (GTZ) GmbH. The authors were assisted in their work by Ria Miclat-Teves in The Philippines, Andrea Gaifami in Mozambique and Nicolas Mastrocola in Ecuador.

The views expressed in this book are those of the authors and do not necessarily reflect those of other individuals or institutions.

1
Introduction

Seed Supply in Developing Countries—an Overview

The potential and quality of seed is extremely important in the complex, diverse and risky areas of the developing world where there is little access to other techniques and technologies to increase agricultural productivity. This is because, in these areas, farmers are operating in marginal and variable environments which are dislocated from national market infrastructures. Much development attention is now focused on these communities, particularly by people-centred organisations like NGOs. This is both for philanthropic reasons (many farmers in these areas are living on the poverty line and have apparently been bypassed by most development activity) and because, for the health of national economies in the long run, these farmers must be brought into the mainstream of the economy.

But national seed systems and associated plant breeders have had little success in meeting the needs of these farmers. Frequently, such systems do not operate in remote, poor areas. Where they do operate, the hybrids and other specialised seeds which they sell are often of little relevance, since they are tailored to yield well in the controlled conditions of areas with higher potential. And, particularly where seed distribution is the responsibility of over-stretched government departments or nationalised seed companies, seeds are often in poor condition when they arrive and delivered too late in the season to be useful. Perhaps most important, few government agencies have the time or resources to find out what types of seed would fit into farming systems in these areas. Even if they did, they would be unlikely to have the resources to meet the multiplicity of needs.

Such organisations, be they government departments or, as is increasingly the case, subsidiaries of multinational seed and agricultural chemical companies, offer the potential to bring to farmers the benefits of high-technology plant breeding which is possible only in the controlled conditions of laboratories and specialised seed farms, and to capitalise on economies of scale. However, they do not provide the only approach to helping farmers to get the most from their seed inputs: seed production and distribution does not have to be organised through big, nationwide institutions.

No farmers buy in all their seed requirements every year: the seed produced by national seed systems usually represents no more than 20 per cent of the total amount planted in developing countries. This is the case both in the highly capital-intensive, input-dependent agriculture of the North and in the low external input farming systems typical of much of the South. Even the seed companies and the government agricultural extension services, which tend to be over-optimistic about potential seed sales, rarely assume farmers will buy new seed more frequently than once in four years. Farmers are not dependent on national seed systems: this would be an unnecessary expense.

There has been little research into farmers' knowledge of the selection, cleaning, treatment and storage of seed ('seed care') in developing countries. The limited work that has been carried out has found communities and individuals with a great deal of skill in using locally developed techniques, fine-tuned to a considerable degree over time. Such techniques apply not only to the more straight-forward processes of cleaning and storing different kinds of seeds, but also to evaluating and selecting strains that have desirable characteristics and maintaining and developing these strains over the years. These on-farm seed care techniques have proven effective in terms of keeping farmers' varieties true to type and keeping seed clean and in good condition for germination. In fact, seed from many farmer-managed seed systems is at least as good and in some cases better than seed from national seed systems.

At the same time, there is increasing evidence that national seed systems in the developing world not only face difficulties in meeting the seed needs of small farmers outside high potential areas, but are also very costly. As a result, many development workers are now asking whether it would not be more effective, and cheaper, to put money into supporting local level seed supply instead. Governments too are increasingly taking this line. And donors are beginning to recognise the existence of 'dual' seed systems, made up of the national, formal sector and the local, farmer-managed sector, and the need to support the latter as well as the former.

This is a radically different strategy from that followed in many developing countries to date. Some of the main differences are in the kinds of agencies that need to be involved and in how their operations need to be organised, both internally and in relation to the farming communities which they support. For some time now, rural development workers have believed that NGOs are more appropriate for this strategy than government departments or commercial companies. Not only do government departments have attendant bureaucratic obligations, but they are operating at reduced levels of services (especially in more difficult areas) under the pressure of Structural Adjustment Programmes. Commercial companies focus on better-resourced farmers in higher potential areas because they are more profitable.

A number of NGOs have already become involved in seed activities. However, the experiences gained from these initiatives have not yet been analysed collectively. It is to fill this gap that this book has been written: to find out whether this approach can create a new *people-centred seed strategy* for supporting local seed systems outside the high potential agricultural areas of the developing world.

Methodology of the Study

The approach to the research for the book was debated in some detail at the start of the study. This was an interactive process involving staff from many NGOs with seeds activities, as well as the research team. It included an initial desk review of the many relevant NGO reports and other unpublished documents that have accumulated at ODI; correspondence with ODI contacts amongst agencies overseas; and interviews with agencies based in the UK and mainland Europe. The aim was to provide answers to questions identified as important by NGOs.

It was established that data was needed in two main areas. The first concerns how well individual agencies perform in relation to a given set of seed system performance indicators. These are outlined in Chapter 7 (see also Cromwell, Friis-Hansen and Turner (1992) which provides a detailed methodology for seed sector performance assessment). The second data set concerns the institutional performance of NGOs compared to other development agencies: in particular, the cost of seed produced in NGO projects and programmes and the sustainability of the seed systems set in place by NGOs. This is defined in Chapter 7. Emphasis was placed on obtaining the comments and opinions of the farmers and government services that the NGOs work with.

The choice of agencies to form case studies was an iterative process. Initially, a list of selection criteria was drawn up. All the case studies were chosen to represent agencies working with local seed systems and farmer communities at a practical level, with food staple crops and with a reasonably well-documented project experience. In addition to this, case studies were chosen to represent: different types of agencies (North-based NGOs, South-based NGOs, donor agencies); a range of development strategies; different types of seeds (landraces, HYVs, etc.); different agro-ecosystems; different macro-economic and policy contexts. Given that there were relatively few agencies which fulfilled these criteria, the final selection was based mainly on pragmatic considerations: could the research team get access to the level of information required? The final selection of 18 case studies is given in Figure 1.1. One third of the agencies reviewed are North-based NGOs (ACORD, OXFAM, Concern, MCC and SCF[US]); one third are North-based NGOs supporting seed activities operated by local organisations (ActionAid, CIC, ASS, GRAIN and RAFI); a quarter are multi- and bi-lateral donors or technical support agencies, supporting local seed systems through government projects (FAO, PPS, KHAP); and the remaining two (CESA and MIND) are South-based NGOs.

For each study, information was collected about the agricultural and policy context of the country, the background of the agency reviewed, its organisation and management of seed activities, the costs of its seed activities, and its current performance and its plans for the future. Information was collected in three different ways. Members of the research team worked in-country in The Gambia (three weeks) and in Nepal (seven weeks). In-country case studies

Figure 1.1: Map Showing Location of Case Studies

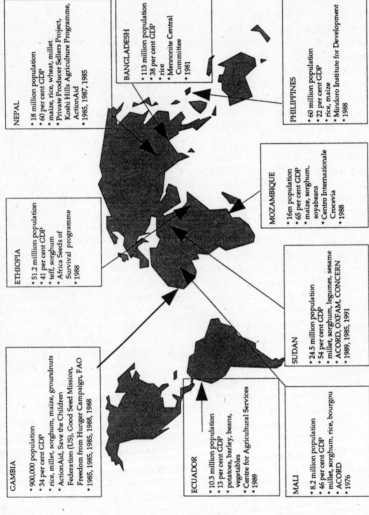

GAMBIA
* 900,000 population
* 34 per cent GDP
* rice, millet, sorghum, maize, groundnuts
* ActionAid, Save the Children Federation (US), Good Seed Mission, Freedom from Hunger Campaign, FAO
* 1985, 1985, 1985, 1988, 1988

ECUADOR
* 10.3 million population
* 13 per cent GDP
* potatoes, barley, beans, vegetables
* Centre for Agricultural Services
* 1989

MALI
* 8.2 million population
* 46 per cent GDP
* millet, sorghum, rice, bourgou
* ACORD
* 1976

ETHIOPIA
* 51.2 million population
* 41 per cent GDP
* teff, sorghum
* Africa Seeds of Survival programme
* 1988

SUDAN
* 24.5 million population
* 54 per cent GDP
* millet, sorghum, legumes, sesame
* ACORD, OXFAM, CONCERN
* 1989, 1985, 1991

NEPAL
* 18 milllion population
* 60 per cent GDP
* maize, rice, wheat, millet
* Private Producer Sellers Project, Koshi Hills Agriculture Programme, ActionAid
* 1985, 1987, 1985

BANGLADESH
* 113 million population
* 38 per cent GDP
* rice
* Mennorite Central Committee
* 1981

MOZAMBIQUE
* 16m population
* 65 per cent GDP
* maize, sorghum, soyabeans
* Centro Internazionale Crocevia
* 1988

PHILIPPINES
* 60 million population
* 22 per cent GDP
* rice, maize
* Mindoro Institute for Development
* 1988

Sources: Case studies; *World Development Report* 1992; *World Bank Tables* 1989-90.

Notes: Figures are latest available; GDP percentage = agriculture as percentage of GDP

Crops = main small farmer crops; agencies = agencies included in case studies; date = year seed project started.

were provided by local collaborators in Ecuador, Mozambique, Bangladesh and The Philippines. Agency documents, supplemented by interviews with agency staff, were collected and analysed in the UK for the Sudan and Ethiopia case studies and for GRAIN and RAFI.

Seed Systems—Definitions

National seed systems are frameworks of institutions linked together through their involvement with the multiplication, processing and distribution of seed. They include government departments, parastatals and private commercial companies (the formal sector) and farmers, their associations and NGOs working with them (local seed systems) (Cromwell, Friis-Hansen and Turner, 1992). The formal sector is bureaucratically organised and generally not location-specific in its operations. The farmer-managed seed systems with which the informal sector works operate mainly within individual communities and tend to be informal or semi-structured in their organisation, changing between locations and over time (Cromwell, 1990).

Seed systems include not only institutions directly involved in producing and selling seed but also those involved in support activities such as agricultural research and extension, seed quality control and certification, etc. Activities range from plant breeding, variety evaluation and release, through seed multiplication, quality control and processing, to distribution both of new varieties for the first time and of regular supplies of fresh seed. The strong two-way links between the different institutions and activities are an important feature of seed systems. This is illustrated in Figure 1.2.

There are differences in what constitutes 'seed' for different types of plants and in the nature of what is supplied as seed by different seed systems. The grains produced by cereal plants, which are typically thought of as seed, are one example. Another are the nuts and beans produced by legumes. Both these types are known as 'true seed'. Other types of planting material are the tubers of potatoes, cassava and other root crops that are an important source of food for many farming communities in developing countries. Some basic information about seed technology and how it affects seed activities is given in Appendix 1.

The seed produced by the formal agricultural research system and the organised seed sector is often called 'improved' or 'quality' seed or 'HYV', referring to the high-yielding varieties that formed the basis for the Green Revolution. However, the implication of these labels—that formal sector seed is better than seed that farmers can themselves produce—is not always justified. In this book, we follow the more recent convention of calling this seed 'modern variety' (MV), referring to the valid distinction that it has been produced using breeding principles only possible in the high-technology systems of the formal sector, but making no value judgement about its relative merits.

The seed produced by farmers themselves we call 'farmers' variety' (FV) or

Diagram 1.2:   The seed sector - a framework approach

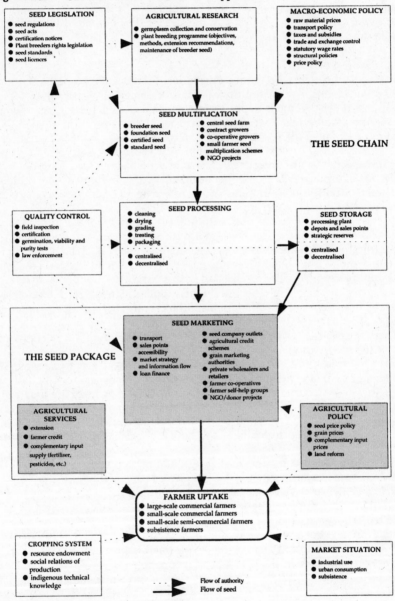

SEED LEGISLATION
- seed regulations
- seed acts
- certification notices
- Plant breeders rights legislation
- seed standards
- seed licences

AGRICULTURAL RESEARCH
- germplasm collection and conservation
- plant breeding programme (objectives, methods, extension recommendations, maintenance of breeder seed)

MACRO-ECONOMIC POLICY
- raw material prices
- transport policy
- taxes and subsidies
- trade and exchange control
- statutory wage rates
- structural policies
- price policy

SEED MULTIPLICATION
- breeder seed
- foundation seed
- certified seed
- standard seed
- central seed farm
- contract growers
- co-operative growers
- small farmer seed multiplication schemes
- NGO projects

THE SEED CHAIN

QUALITY CONTROL
- field inspection
- certification
- germination, viability and purity tests
- law enforcement

SEED PROCESSING
- cleaning
- drying
- grading
- treating
- packaging
- centralised
- decentralised

SEED STORAGE
- processing plant
- depots and sales points
- strategic reserves
- centralised
- decentralised

THE SEED PACKAGE

SEED MARKETING
- transport
- sales points accessibility
- market strategy and information flow
- loan finance
- seed company outlets
- agricultural credit schemes
- grain marketing authorities
- private wholesalers and retailers
- farmer co-operatives
- farmer self-help groups
- NGO/donor projects

AGRICULTURAL SERVICES
- extension
- farmer credit
- complementary input supply (fertiliser, pesticides, etc.)

AGRICULTURAL POLICY
- seed price policy
- grain prices
- complementary input prices
- land reform

FARMER UPTAKE
- large-scale commercial farmers
- small-scale commercial farmers
- small-scale semi-commercial farmers
- subsistence farmers

CROPPING SYSTEM
- resource endowment
- social relations of production
- indigenous technical knowledge

MARKET SITUATION
- industrial use
- urban consumption
- subsistence

Flow of authority
Flow of seed

Source: Cromwell, Friis-Hansen and Turner, 1992.

'local variety'. This includes both indigenous plant genetic resources (landraces) and seeds that have had elements of exotic material incorporated either by accident (for example, through uncontrolled cross-pollination) or deliberately, using farmers' conventional breeding techniques of mass selection on the basis of visual characteristics. Farmers' varieties are often not distinct (different varieties can share a number of the same characteristics) nor stable (there may be considerable variation in characteristics within the same 'variety'). Thus, a relatively large degree of genetic diversity can be maintained with few farmers' varieties.

Also, it is important to be aware that there is a rich range of 'intermediate' varieties between 'modern varieties' and 'farmers' varieties' rather than a simple dichotomy between ancient local landraces and modern synthesised hybrids. These include what we call enhanced farmers' varieties (referring to improvements in physical seed quality); improved farmers' varieties (referring to improvements through breeding and selection); locally-adapted MVs (also through breeding and selection); and degenerated modern varieties (i.e. MVs that have been maintained locally by farmers after an initial release from the formal sector).

Loss of genetic diversity, or 'genetic erosion' is an increasing concern in many national seed systems. This refers both to the loss of individual genes and to the loss of varieties; both damage the sustainability of agriculture in the long run. It is essential to maintain a broad plant genetic base: the development and use of modern varieties is not necessarily damaging, as long as the MVs that are bred contain this broad genetic base—the danger arises from modern plant breeding methods, which tend to incorporate only a few genes into each new variety.

Fears of genetic erosion primarily concern the *breadth* of the world's plant genetic resources base. However, the *control* of this base is becoming an equally important issue because of developments in the area of intellectual property rights (particularly patents). Concerns about control are most frequently expressed at international level in the work of organisations such as GRAIN and RAFI: individual projects supporting local seed systems are usually less involved in such issues (ICDA, 1989). However, if the current global pressure for patents is successful, increased commercial control of plant genetic resources at the international level may preclude efforts to maintain the breadth of the genetic base at project level.

How This Book is Organised

This book follows on from the work of Cooper, Vellvé and Hobbelink (1992), which explored the technicalities of genetic resources conservation and improvement at the grassroots level. We are concerned to document for NGOs the institutional implications of their work in supporting local seed systems. These are very important, but are often ignored in the face of more immediate concerns about genetic resources conservation. We are especially concerned to move analysis beyond the restrictive notion that in the absence of apparent alternatives NGOs are obliged to replicate existing formal sector methods of organising seed production and distribution. We demonstrate that farmers, governments and NGOs can benefit from multi-institutional approaches tailored to supporting farmer-managed seed systems already operating within communities.

The first part of Chapter 2 synthesises what is known about how farmer-managed seed systems operate. The second part of Chapter 2 sets out the principles espoused by many NGOs in their development work and it draws out aspects of their approaches particularly relevant to seed supply. The third part of Chapter 2 gives an overview of where, how and why NGOs have become involved in seed activities in developing countries.

This sets the scene for the case studies that form Chapters 3 to 6. These document a range of experiences of work with seeds carried out by 16 different agencies in nine countries, and two NGOs working on seeds internationally (see Figure 1.1). Most of the case studies are presented in this format for the first time.

Chapter 7 uses the case studies to assess the effectiveness of the various approaches to supporting local seed systems tried by different types of agency. Following the definition of 'sustainability' in the context of seeds work provided in that chapter, specific attention is paid to technical soundness, organisational and managerial effectiveness and sustainability, and comparative economic costs and benefits.

The issues which recur throughout the book, and on which we draw together findings in Chapter 8, are as follows:

- **farmer-managed seed systems**: how do these systems operate? what are the main problems facing them? how can external support help?
- **NGOs and seeds**: what are the comparative advantages of NGOs' approach to seeds work compared to other types of development agency? what are the weaknesses to guard against?
- **sustainability of local seed systems**: what does 'sustainability' mean in the context of seeds work? what factors influence long-run institutional and economic sustainability? are there trade-offs between them?
- **genetic diversity**: how can global concerns about genetic erosion be translated into local strategies for the conservation and use of a broad plant genetic base? what blend of modern and farmers' varieties is most useful to developing country farmers? what does this imply for plant breeding and variety evaluation methods?
- **governments and donors**: how can support for farmer-managed seed systems be encouraged? what influence does macro-economic policy have? in what areas is co-ordinated global action required?

2
Farmers, NGOs and Seeds

Small Farmers and their Seed Needs in Context

Despite urbanisation, the majority of the world's population still lives in rural areas, and most of these rural dwellers engage in agriculture. In 1990, about 2,400 million people depended primarily on farming and herding for their livelihoods (FAO estimates). The vast bulk of these live in the developing world, and most of them live in small-scale farming households, of which there may be more than 375 million. What do these households typically have in common? Ellis (1988) defines small farmers—technically called 'peasants' in his work—thus:

> 'Farm households, with access to their means of livelihood in land, utilising mainly family labour in farm production, always located in a larger economic system, but fundamentally characterised by partial engagement in markets which tend to function with a high degree of imperfection.' (p.12)

Small farmers typically have access to only small areas of land, frequently less than two hectares: where they have access to large areas in absolute terms, the land is almost invariably of low potential so that its effective area is relatively small. They usually produce a significant amount of their own food, although few are self-sufficient, as well as other important goods and services—housing, fuel, water, some tools, sometimes clothing, etc. Most household labour is devoted to farming, and most of their income comes from farming. However, off-farm activities (paid work, trading, transport) and non-agricultural occupations (crafts, artisan work, food and raw material processing, village services) are important: surveys repeatedly find farmers earning 25 to 50 per cent of their income from these other activities. For most small farmers, economic conditions are precarious and few have the chance to build up wealth and capital. Lack of economic power leaves them with little political power, and they find themselves subordinate—economically, socially, and politically.

That said, small-scale farming takes place under an astonishing heterogeneity of conditions. To simplify, a primary distinction can be made between favoured and marginal areas. The better areas have more fertile soils, dependable rains or irrigation. However, in these more favoured areas, small farmers often have precarious access to land, renting or share-cropping the land from landowners. Cropping in such areas may be intensive, with high labour use and applications of manure and fertiliser, resulting in high yields.

However, the potential profits from the land accrue only in part to the small farmer since the landowner typically takes a significant proportion in rent. Population densities tend to be high (300 per square kilometre and more) and households have access to very small areas. Moreover, population growth exacerbates land hunger as with every generation the farms are subdivided further.

The marginal lands are marred by aridity, low soil fertility, mountainous terrain with shallow, acidic, and stony soils, and are often remote from urban centres. They present a different picture. Here, independent small farmers have more assured access to their land and rarely pay for it, and typically have relatively large areas to farm. Labour is usually the limiting factor of production, and the farming is often low-input, low-output. Agriculture in the marginal lands is frequently closely integrated into the natural environment, making careful use of the biological systems already in place with relatively little disturbance of the ecosystems. That does not, however, imply security: farming is still risky in these lands, no matter how closely it is in tune with local conditions. Drought is the main hazard, but flooding, pests and diseases all constitute severe problems. Consequently small farmers pursue risk-avoiding or risk-mitigating strategies: mixed cropping, multiple enterprises, keeping savings (often in kind), storage of food, investing in social networks of mutual assistance, mobility, etc.

In both the favoured areas and the marginal lands, important changes have occurred in the last few decades and continue to take place. Almost everywhere, and despite substantial out-migration to the towns and cities, the farming population is increasing. This puts pressure on farming systems to intensify production and leads to farm families moving to uncultivated margins (semi-arid tracts, steep slopes and tropical forests) to find new land.

Economic growth, however faltering, has meant the spread of increased opportunities for trade and greater integration in regional and national markets. It has become possible and attractive to sell crops, to buy manufactured farm inputs (tools, fertilisers, pesticides, etc.), and to purchase consumer goods previously unheard of in the countryside (radios, bicycles, flashlights, metal and plastic cooking utensils and all kinds of clothing). This has pushed up the cost of 'subsistence' and created a strong incentive to greater production and sales in the markets for farmers.

New technology has become available. Sometimes it is simple, like better tools (hoes, metal ploughs), sometimes more sophisticated, like tubewells and diesel pumps. Above all, new technology has meant the spread of packages of high-yielding varieties of cereals, together with water, and fertiliser, that were promoted as part of the Green Revolution that occurred from the mid-1960s onwards in Asia. This followed the development of wheat and rice varieties, by the international agricultural research centres, with far higher yields than before given large applications of external inputs. For the favoured areas, the Green Revolution has allowed sustained increases in farm output even in areas which already had relatively high yields. In the marginal lands, however, the Green Revolution package does not work: technical advances have been limited and halting and no such quantum leaps in production have been registered.

In both expanding market opportunities and in spreading new technologies, governments have been active in promoting agricultural development by offering support services and by building infrastructure.

Finally, in some areas, again usually the more marginal, environments are changing under the impact of intensified land use, for example, loss of soil cover and of fuelwood trees. Even the climate may be in flux; long-term changes in rainfall patterns in some parts of the Sahel being the best-known example. This presents serious challenges to farming systems closely integrated into their natural habitat, where the effects on farming can be severe.

The diversity and complexity of their farming systems reflect the heterogeneous conditions under which small farmers in the developing world operate. Although it is possible to make some generalisations about the economic conditions in which small farmers typically find themselves, it is much more difficult to generalise about the mix of crops grown, the cultivation techniques used and the implications of these for agronomic change. Indeed, it is over-generalisations in this area which have caused many of the problems now facing agricultural development programmes attempting to reach small farmers effectively, particularly those in the more marginal environments.

In the more favoured areas of Asia where fertiliser, irrigation and farm labour are available and rice and wheat dominate, plant breeders have found it relatively easy to produce varieties with high potential yields that can be achieved in this type of farming system. Accordingly, the coverage of high yielding modern varieties is now substantial in these areas. The key issues now are: equity effects, especially for those households without access to irrigation and too poor to afford fertiliser; environmental effects of high fertiliser and pesticide applications; and longer-term effects of reduced genetic diversity, as a few highly stable and uniform varieties cover larger and larger areas of land. One of the most obvious negative effects of this is the increased incidence of pest and disease attack. This ties farmers into regular purchase of new varieties (and of chemicals), as the disease resistance of the old ones breaks down.

In contrast, in the more marginal areas, there is no irrigation, labour is in short supply and markets for fertiliser are far away. The farming system is altogether more complex and diverse and the maize, millet, sorghum, cassava and potatoes which dominate high-yielding varieties can be produced less readily. In addition, plant breeders cannot use conventional breeding techniques to great effect for these areas as they require many niche varieties, pest and disease resistance, drought-tolerance and a whole range of other attributes in addition to high potential yield. The uptake of modern varieties in these areas is low and often only for a few niche crops. In these areas, the key issues now are: is the introduction of outside varieties the most appropriate intervention (compared to changes in cultivation techniques, diversification, off-farm income, etc.); and do plant breeding techniques need radical change in order to provide these areas with useful material?

Nonetheless, some broad generalisations can be made about small farmers' seed needs and attitudes to modern varieties and the factors that affect them. These are summarised in Boxes 2.1 and 2.2 respectively.

The Role of Formal Sector Seed Systems

Many reports on individual seed projects and programmes have pointed out the enormous difficulties faced in getting a formal national seed system to provide an effective service to small farmers in marginal, variable environments. Two recent studies, by Cromwell, Friis-Hansen and Turner (1992) and by Groosman (1991), have confirmed this picture on a wide scale in developing countries.

The formal sector finds it most profitable to produce seed of open-pollinated crops, especially hybrids, and of vegetable crops. It also concentrates on seed of crops with high sowing rates (more is bought) and multiplication factors (more can be produced quickly).[1] It therefore tends to ignore certain crops that are important to small farmers, such as self-pollinated cereals and legumes. At any one time it provides a limited number of widely adapted varieties. These are not generally varieties that small farmers can make use of, because the most useful varieties to such farmers tend to be less profitable for seed companies and to face stiff competition from farm-saved seed. The methods and orientation of plant breeders within the formal agricultural research system—not only the objectives of the seed companies themselves—have an important influence on the relevance of the new varieties made available.

The formal sector often faces difficulties with transport and storage, meaning that seed delivery is usually more of a problem than seed production. This is exacerbated for the small farm areas where markets are limited. Thus, over much of the developing world, the formal seed sector does not reach small farmers, either because it does not exist (as is the case in much of Africa) or because although it does exist, it is not oriented towards serving small farmers. In seed institutions in the public sector, inefficiency caused by over-bureaucratisation is a problem. Some national seed projects suffer from the additional problem of being reliant on high technology seed production and processing systems which are costly to maintain in the long run.

Thus in terms of both the product required and the distribution system, the formal sector often cannot provide an effective seed service in the marginal, variable environments of the developing world, except for certain niche markets.

Added to this, many farmers have misconceptions about formal sector seed: they do not believe that it has superior genetic potential unless they see it growing; they do not believe it is of better physical quality, because few

---

1. These terms are explained in Appendix 1.

Box 2.1: Small farmers' seed needs—a summary

- **Seed varieties**: each farmer uses a large number of varieties of each crop and in addition many look for intra-varietal variation rather than uniformity and stability. Non-yield attributes (taste, storability, straw yield, etc.) are important. Demand for local farmers' varieties can be as strong or stronger than for exotic modern varieties. The specific characteristics required in any given situation will depend on the function of each crop in the local farming system both agronomically and economically.

- **Seed quality**: seed that meets formal ISTA quality standards may be unnecessary but seed has to be of proven and reliable physiological quality for it to be demanded by small farmers. Providing seed of proven superior physiological quality to what farmers are currently saving themselves can be more valued than providing new genetic material.

- **Quantity of seed**: the quantity of seed an individual farmer requires depends on prevailing seed replacement and sowing rates (which are not necessarily the same as those recommended by the research and extension services). The quantity required each year for a given crop may be extremely small and farmers often want to be able to buy seed in small packets.

- **Timeliness of seed delivery**: seed from outside must arrive in good time for farmers' preparation and planting timetable. Again, this may differ from the one recommended by the agricultural services. This is one of the most important small farmers' seed needs—because their labour-intensive production methods mean planting at the right time is critical, but they do not have sufficient cash to buy seed in advance and store it.

- **Accessibility of seed delivery points**: farmers are often relatively indifferent to the distance that has to be travelled to fetch seed, so long as it is known to be of good quality. But they can have strong preferences for different kinds of seed delivery points (government crop authority, private seed company, general trader, other farmers, etc.), which vary between crop.

- **Retail seed prices**: all but the poorest and most insecure families are relatively indifferent to the price charged for seed, so long as it is of proven good quality, because seed purchases make up a relatively small proportion of total production costs. However, many families find it difficult to pay cash for seed and will seek alternative methods (labouring, in-kind loans, etc.)

- **Support services**: the potential benefit of using a certain variety of seed often depends on the use of fertiliser or some other external input. In this case, there will be no demand for the seed unless there is a service providing these complementary inputs.

Source: Cromwell, Friis-Hansen and Turner, 1992.

Box 2.2: Area coverage of modern varieties

Agriwal (in CIAT, 1982) points out that the area coverage of modern varieties tends to be greater for crops for which hybrid varieties exist and for which the grain is not the end product. Thus, in *El Salvador*, modern varieties of *maize* are now estimated to cover 66 per cent of the cropped area (Puentes in CIAT, 1982).

Exogenous changes can have an important effect on the area coverage of modern varieties. Wiggins (1992) records that in areas of *The Gambia* experiencing a long-term decline in the amount and reliability of rainfall, modern *rice* varieties have replaced farmers' long-season varieties and now cover 80 per cent of the area.

There is an important distinction between the area coverage of modern varieties, which can be relatively high for certain crops in certain areas, and the proportion of seed which is purchased fresh every year (the seed replacement rate). In the *Andean region* of Latin America, for example, 50–80 per cent of the *potato* area is estimated to be planted with modern varieties but only 3–10 per cent of this is bought in as fresh seed every year (Monares in CIAT, 1982). In contrast, in the *Great Lakes region* of East Africa, Sperling *et al.* (1992) have found that although 40 per cent of farmers obtain some *bean* seed off-farm, only 10–30 per cent use new varieties.

Numerous different sources agree that overall some 80 per cent of the cropped area in developing countries is covered by seed which has not passed through any marketing channel (Delouche in CIAT, 1982; Agriwal in CIAT, 1982; Osborn, 1990; Bal and Douglas, 1992).

There can be considerable variation in the use of purchased seed between areas. For example, in *Colombia*, in one area up to 80 per cent of *maize* seed is purchased every year whereas in another only 10 per cent is purchased (Velasquez in CIAT, 1982). There is also variation between crops: the same data show that whilst up to 80 per cent of *maize* seed is purchased every year, only 50 per cent of *bean* seed is bought in.

---

quality attributes are visible in the seed itself; and they often consider it expensive, as they do not take into account the opportunity cost of using saved grain as seed, nor do they measure seed costs against total variable production costs, including labour.

Undoubtedly, for a number of crops, saving seed rather than buying it in from the formal sector is a rational agronomic and economic strategy for small farmers. This is sometimes because of factors specific to particular crops, such as self-pollinated crops, which can be maintained easily on-farm. Or it can be the result of economic factors. For example, seed for domestic food crops tends to be saved on-farm because such crops produce little cash income to spend on purchased seed. Similarly, seed tends to be saved on-farm for crops where purchased seed would form a high proportion of total production costs because sowing rates are high, such as rice and some legumes.

However, saving seed on-farm is *not* a rational strategy in circumstances where crops do not seed in tropical climates (for example, many exotic vegetables), where seed-borne diseases are a major problem (often the case for beans) and where seed deteriorates rapidly in hot, humid conditions and has to be stored over seasons where these conditions prevail (as is the case with wheat in parts of South East Asia and with soyabeans). There is also a negative trade-off between saving seed and maximising production where grain crops are inter-planted or stagger-planted as there is a substantial reduction in total production in these circumstances if the crop is grown pure stand for seed.

Farmer-Managed Seed Systems

Despite the fact that traditional seed systems are often much more important to small farmers than the formal seed sector, virtually no attention has been given to the seed sourcing and seed care practices of farmers themselves. There is a need to compile, interpret and use the data that is beginning to accrue in many countries; we present here a summary of the information available to us. The typical small farmer seed production process is outlined in Box 2.3.

Plant Breeding

Farmers use not only landraces and modern varieties from the formal sector, but also varieties that farmers have developed themselves. Before scientific plant breeding began in the late nineteenth century, the genetic improvement of crops depended entirely on farmers' selection from local material, using visual characteristics such as yield, grain size and colour. Thus farmers have for many centuries been actively involved in plant breeding and their breeding skills are highly developed.

The most straight-forward technique is mass selection, but there is evidence that farmers carry out controlled crossing (Montecinos and Altieri in Cooper *et al.*, 1992). Carefully documented variety performance records have been found in a number of farming communities (for example, in Ethiopia as described by Mooney in Cooper *et al.*, 1992). The ASS project in Ethiopia estimates that yield increases of 3.5 per cent per year can be obtained from on-farm selection from farmers' local varieties of crops such as teff, sorghum and millet (USC, 1988). And there are a number of other documented examples of farmers' success in variety development. These include the experience of the MASIPAG coalition in The Philippines, which has been supporting farmers' groups developing low-input alternatives to the formal sector's high-yielding rice varieties: in just seven years, following nationwide collections of potential germplasm, over 100 lines have been selected for further crossing (Villegas, pers. comm.).

The community-controlled seed selection and maintenance practised in Tigray, Ethiopia, has even been put forward as a means for farmers to be eligible to hold intellectual property rights, in competition with multinational agro-chemical companies (Berg, 1992). When seed banks were established in

---

Box 2.3: Farmer-managed seed production process

- **Field production**: many small farmers rogue their growing crops by hand to remove diseased plants, and, less commonly, remove off-type plants as well. Farmers usually do mass selection on basis of visual appearance of individual grains rather than plants. This is usually, although not always, done in-field pre-harvest.

- **Harvest**: crops are harvested by hand so mechanical damage to the seed and contamination with seeds and other inert material are avoided.

- **Cleaning**: after harvest, crops are often threshed and cleaned by hand, again limiting damage and contamination.

- **Drying**: crops are usually dried in the sun, which can reduce moisture content to satisfactory levels—although there is some danger of scorching and killing seeds if the crop is left in the direct sun for long periods.

- **Storage**: considerable care is often taken in the storage of seeds: local insecticides and fungicides (for example, eucalyptus leaves, sand, ash, neem) can be added to the crop, which is then placed in special sealed containers which are themselves stored in places, such as above the fireplace, best suited to keeping the seed pest- and disease-free and viable. In damp climates, seed is often removed from store and re-dried a number of times during the course of the storage period.

- **Conditioning**: it is less common for small farmers to do any kind of germination testing prior to planting the stored seed and there is no documentation of any traditional pre-germination seed treatments being applied. However, in certain areas, where the incidence of pre-germination pest and fungus attack is high, and modern chemical seed treatments are cheaply available (for example, in Eastern Kenya, parts of Mali and parts of Eastern Sudan), these are now widely used on saved seed.

Source: Delouche in CIAT, 1982.

Tigray in response to the droughts and wars of the late 1980s, it was found—contrary to expectations—that farmers had finely-tuned seed selection skills which allowed them to maintain a large diversity of varieties of different crops, including small-seeded ones such as teff and millet.

Where farmers find it difficult to select varieties for certain characteristics, they often develop alternative techniques for dealing with the problem. A good example of this is the technique of planting mixes of large numbers of individual bean varieties in single plots, which has been well documented in

East Africa. This technique allows farmers to overcome the difficulty of selecting individual bean varieties with resistance to the wide range of prevalent pests and diseases: instead, they can be sure that, whatever combination of pests or diseases attacks their crop in any one season, there will be a sufficient proportion of plants resistant to that combination. The value of this technique is being recognised by some formal sector bean breeders and seed producers. Instead of trying to produce bean varieties to replace traditional mixes, breeders are now aiming to produce varieties that can be incorporated into existing mixes. This new approach is well documented in, for example, the reports of Michigan University CRSP (MSU, 1987).

It is a fallacy that the risk-averseness of small farmers in marginal environments means that they do not experiment; risk-averseness does not equate with conservatism and small farmers in risky environments often pursue the economically logical strategy of conducting small-scale experiments with a wide variety of planting material.

Seed Storage

The neglect of on-farm seed storage practices and potential is all the more surprising given that the importance of improving on-farm storage of grain (as opposed to seed) has been recognised for some considerable time. The limited evidence that has been collected (by, for example, CIAT in Latin America and the Great Lakes region of East Africa, and by Pakhribas Agricultural Centre in Eastern Nepal) suggests that there is a considerable amount of indigenous technical knowledge within farming communities about solutions to local storage problems using locally-available materials. Salazar (in Cooper *et al.*, 1992) states that the experience has been similar in South East Asia for various important food crops.

Nonetheless, losses in store can be large, often exacerbated by climate. For example, in Nepal losses of maize stored on farm at altitudes where weevils are a problem have been estimated at up to 50 per cent of the stored crop, whereas at higher altitudes, losses fall to practically zero (Cromwell, Gurung and Urben, 1992). Losses of wheat in Asian farming systems where it is stored through the wet, rice-producing season have also been recorded as significant.

Where simple changes in storage structures have been introduced, there have often been very significant improvements in the quantity of usable material available to farmers at the end of the storage season. Some of the most spectacular improvements have been obtained in potato storage, for example with the rustic potato store design developed by the National Potato Development Programme in Nepal and a similar design produced by the National Tuber Programme in Colombia. But there have also been improvements in grain seed storage, such as the low-cost metal seed bins distributed by the Rural Save Grain Project in Nepal (NSB, 1990) and the community seed bank kit promoted by Rural Advancement Foundation International (RAFI, 1986).

Seed Quality

Farmers' indigenous seed care skills and technologies appear consistently to produce seed of equal or better quality to that produced by the formal seed sector. They may indeed have an inherent advantage in doing so, because it is much easier to carry out the special procedures required when the quantities are small. This has been documented for maize in Latin America (Delouche in CIAT, 1992), for beans in East Africa (for example, CIAT, 1991) and for other crops elsewhere in the developing world (Linnemann and de Bruyn, 1987). The CIAT Great Lakes Regional Programme in East Africa recently conducted experiments to measure the comparative quality of bean seed saved by small farmers and seed obtained from the local agricultural research station and found 'no statistical differences . . . in terms of vigour, emergence and yield' (CIAT, 1992:4). The On-Farm Seed Production Project which operates in Senegal and The Gambia found that, for self-pollinated crops like rice, varietal purity of farm-saved seed is high (Osborn, 1990).

Some development programmes, such as Pakhribas Agricultural Centre in Nepal and CIAT in Latin America, have drawn up special seed production guidelines geared to small farmers' circumstances. However, many others still promote practices that are unrealistic: for example, requiring maize to be sized (for mechanical planting) and groundnuts to be shelled. This is because field production and laboratory testing standards have often been copied from seed quality control regulations developed for large-scale, capital-intensive agriculture in the temperate climates of developed countries. These are necessary where seed production and use are physically separate, because many aspects of seed quality are not visible. However, they are not necessary to the same degree in farmer-managed seed systems, where seed users see the seed growing, or know the growers personally. Furthermore, formal seed certification is only a means of guaranteeing quality which is not visible: seed often deteriorates or is mixed when bags are split in store after certification and neighbour certification (the informal validation of seed quality by neighbours' visual supervision 'over the fence') is just as valid where seed does not leave the community where it is produced.

Formal sector standards are thus unnecessary and unrealistic for farmer-managed seed systems. The need now is to collect sufficient information about small farmers' seed needs to be able to develop seed quality standards appropriate to these needs. Too often, farmers are using low-cost systems of cultivation, harvest and storage which produce good results but formal sector researchers and extension workers 'do not know about these practices and insist upon more sophisticated and costly methods' (de Queiroz in CIAT, 1982).

Small Farmer Seed Care: Possible Improvements

While formal seed certification is often not necessary, there are some modifications that can be made to small farmers' typical seed care practices which would increase the quantity of seed saved and also the range of crops and varieties.

Existing practices for traditional crops in the farming system can be improved. For example, for certain crops in certain environments, the varieties already in use are well adapted to local conditions and it is relatively easy to maintain varietal purity on-farm but maintaining physical quality (seed health, germination capacity, etc.) is difficult. In this situation, the main requirement is for help with improving the quality of existing varieties rather than for the introduction of new varieties. This is the case, for example, for soyabeans in the hot humid conditions of Lakeshore Malawi (Cromwell and Zambezi, 1992); and for wheat stored over the humid season in Nepal (Cromwell, Gurung and Urben, 1992).

But improvements in seed care practices are particularly important where new crops and varieties are being introduced, either for increased productivity or diversification or in response to climatic change. This is the case, for example, with beans in the Great Lakes Region of East Africa, where physical quality is maintained to a high standard on-farm and the need now is for new genetic material (Sperling *et al*, 1992). Such modifications include:

- *production practices*: where increasing the quantity of seed produced is important, planting seed crops at lower densities than grain crops can increase the final seed yield. Keeping seed plots separate from food plots can reinforce the distinction in farmers' minds that seed plots need slightly more attention for maximum results. For seed plots of crops prone to pest and disease attack when grown in pure stand, for example beans, it may be worthwhile to introduce special pest and disease control practices, either biological methods or chemical sprays;
- *varietal purity*: for self-pollinated crops, simple guidelines for how to prevent mixing of seed of different varieties during harvest, on drying floors and in storage; for cross-pollinated crops, training in how to isolate seed plots of different varieties either by distance or in time (for example, by staggered planting, as has been introduced with some success for open-pollinated maize varieties in the hills of Nepal); and training in how to rogue off-types and weeds in crop stands as well as diseased plants. Encouraging in-field selection of seed during the growing season, rather than selection after harvest, allows farmers to select for characteristics such as plant type as well as grain size and colour;
- *germination*: locally-adapted training in the best time to harvest in order to maximise germination (this may involve early harvesting, before the rest of the crop, or late harvesting after rains have finished, depending on the area and variety), in rapid drying techniques and in subsequent storage. In some situations, it may be relevant to introduce simple pre-planting germination tests and methods for breaking dormancy in recalcitrant seeds;
- *seed health*: encouraging the rotation of plots used for seed production in order to minimise the build-up of pests and diseases; training in the

recognition and removal of diseased grains, in simple insect control techniques in storage, and in simple seed health tests (for example, floating seeds in water to identify light-weight, non-viable seeds).

CIAT has found that in various farming systems in Latin America, significant improvements in the quantity of usable farm-saved bean seed can be obtained simply by modifying the traditional threshing technique of pounding beans with sticks, which has been resulting in considerable mechanical damage (Voysest in CIAT, 1982). Similarly, the Near East Foundation has found in parts of Mali that farmers can no longer sow sorghum off the head because of climatic changes. Information about simple seed storage methods is of considerable help (NEF, 1988). ACORD's seed bank programme in Eastern Sudan has found that having introduced simple techniques for improving the physiological quality of seed, local varieties performed as well as, if not better than, the so-called improved varieties (Renton, *pers. comm.*).

Small Farmers' Seed Sourcing

In most parts of the developing world, the choice farmers have in obtaining their seed supplies is not simply between saving seed on-farm or buying it from one of the national seed programmes: there is a whole range of traditional seed distribution mechanisms that operate within farming communities (Cromwell, 1990). In fact, seed sourcing patterns are often relatively complex, with farmers obtaining different crops and varieties from different sources at different times.

Within small farm communities, there appear to be four different categories of seed users (Sperling *et al.*, 1992; Cromwell and Zambezi, 1992): those that are *seed secure*; those that source seed off-farm from time to time out of *choice*; those that source seed off-farm from time to time out of *necessity* (usually due to some kind of domestic disaster); and those that have a chronically *insecure* seed supply and consistently need to source seed off-farm. The proportion of households falling into each of these categories varies from place to place and over time; also, households can be in different categories for different crops.

One of the important distinctions, as was illustrated in Box 2.2, is between sources used to obtain a new variety for the first time and those used to obtain fresh seed of varieties already in use. This latter (the rate of seed replacement) is usually much lower than is recommended by formal sector seed technologists (see Appendix Table 1.1). Data from Nepal, for example, suggest that farmers typically replace wheat seed every seven years, maize seed (open-pollinated) every ten years and rice seed only once in twenty years (Rajbhandary *et al.*, 1987). Monares (in CIAT 1982) calculates that average replacement times for seed potato in the Andean region of Latin America is eight to ten years. In Malawi, 75 per cent of households growing soyabeans and over 40 per cent of households growing beans, replace seed less frequently than every five years (Cromwell and Zambezi, 1992).

In Ethiopia a recent seed survey found that between 25 per cent and 50 per cent of small farm households borrow or buy seeds every year but most transactions take place between neighbours and relatives; farmers say they prefer this system because they can see the crop stands from which the seed is taken (Singh, 1990). In Malawi, two thirds of all bean seed used by small farmers is obtained from neighbours, relatives or other local sources (Cromwell and Zambezi, 1992), whilst in the Great Lakes region of East Africa 75 to 85 per cent of bean seed is originally obtained from relatives (Sperling *et al.*, 1992). In Nepal, 94 per cent of farmers in the Koshi Hills who had taken up a new rice variety, *Pokhreli Masino*, had obtained seed from other farmers rather than from the formal sector (Green, 1987). In the Punjab in Pakistan, two thirds of farmers sourcing wheat seed off-farm approach neighbouring farmers rather than going to the local Punjab Seed Corporation depot (Heisey, 1990).

In-kind seed loans are important in many areas. Non-cash transactions are an important means of giving a wide range of socio-economic groups access to seed: the need to pay cash is cited as a disincentive to the use of formal sector seed in Nepal (Cromwell, Gurung and Urben, 1992), among other countries.

One of the main advantages of community mechanisms is that they are not dependent on transport and communications infrastructure. However, farmers often travel a long way to source seed through traditional community systems. Recent survey data indicates farmers travelling 30 kms in Malawi for beans and five days' walk in Nepal for potatoes (Cromwell and Zambezi, 1992; Cromwell, Gurung and Urben, 1992).

However, there is no evidence that individual farmers set themselves up permanently as large-scale seed producers for sale within the local community. Rather it appears that individuals, who may change from year to year, are approached by other members of the community because they are seen to have a good stand of crops growing or they have planted a new variety which appears to be performing well. The exception to this is where individuals with some kind of traditional status within the community (village headman, large landowners, etc.) are consistently approached by poorer households in pattern of traditional obligations of patronage (as has been documented in Mali and Malawi, for example). Some seed diffusers also display 'a personal commitment and interest in promoting development in their community' (Green, 1987:23): they are not simply the richer farmers or those who have access to new varieties first.

Traditional seed diffusion mechanisms have five key features distinguishing them from the seed distribution systems of the formal sector:

• they are **traditional**: not necessarily static over time in the way they operate but well-established and often elaborate structures, based on and developing out of the traditional channels of information and exchange existing within the community;

- they are **informal** or semi-structured in their organisation, changing between locations and over time, and not subject to the same rigidities of ownership and control as formal sector organisations;
- they operate mainly, although not exclusively, at the **community level**, although lines of supply may extend over a relatively wide geographical area;
- a wide range of **exchange mechanisms** are used to transfer seed between individuals and households, including barter and transfers based on social obligations;
- the individual **quantities** of seed thus exchanged are often very small compared to the amounts formal sector organisations typically deal in.

There is considerable scope for building on traditional farmer-managed seed systems instead of replacing them with outside systems that are likely to be less equitable and less accessible, both physically and economically, and unable to supply the type of locally-adapted varieties that the majority of small farmers want. Various options are explored in Chapters 7 and 8.

NGOs in Rural Development
The positive contribution NGOs can make to sustainable grassroots development is being increasingly recognised by developing country governments and donors. By 1987, for example, the UK was channelling nearly US\$30 million of aid through NGOs each year; the equivalent for the EEC was US\$210 million (ODI, 1988). Accordingly, the number of development NGOs has increased dramatically during the last decade and now there are several thousand based in the OECD countries alone.

Types of NGOs and their Objectives
Within the overall definition of NGOs as 'any organisation that is operationally distinct from government' (ODI, 1988:1), there is considerable variation in their historical origins, their size and significance compared to other agencies and the strength of their links with these other agencies (Farrington and Biggs, 1990). They can be grouped loosely into the following categories:

- **service provision**, ranging from short-term relief to long-term development;
- **organisation-building**, working with local communities to identify problems and to solve them locally, and so to empower them;
- **support and advocacy**, including lobbying national and international policy-makers and providing back-up services such as research and policy analysis and information exchange for smaller organisations;
- **volunteer agencies**, which are primarily geared to providing volunteer technical assistance to other projects and programmes.

It is possible to distinguish two main types of NGO (Bebbington and Farrington, 1993):

- **grassroots organisations**, which seek to promote the welfare of members on a local scale through an agreed set of activities;
- **non-governmental organisations**, both Northern- and Southern-based, which share the philanthropic orientation of grassroots organisations but tend to be more formally institutionalised and are not normally membership organisations. They vary in size and in mode of operation, some implementing projects directly but many working through local NGOs or grassroots organisations or as support organisations for them.

In this book, we refer to both types as NGOs.

NGOs' Strengths and Weaknesses[2]

NGOs are commonly perceived to have various advantages in working for rural development with small farmers in marginal, variable environments, compared to formal sector agricultural research and extension institutions. They can respond to needs quickly and they are often more participatory. They often work with disadvantaged groups in disadvantaged areas. They are independent and can be flexible in their choice of work, information sources, communication methods, clientele and organisational structure. They can adopt an integrated approach to programmes, which includes attention to the institutional and economic context as well as to technical factors.

In addition to the specific advantages of the NGO approach to rural development, there is growing awareness of the substantial unexploited complementarities between agencies within and outside the public sector (Farrington and Biggs, 1990; Merrill-Sands and Kaimowitz, 1990) and of the need for multi-institutional approaches to agricultural technology generation and dissemination.

This has resulted in part from developments in understanding of the process of innovation and change. The 'central source' model of agricultural development has been rejected and NGOs are seen as having an important role as bridging organisations, bringing together diverse actors in economic and social development (Brown, 1991). This role has been accentuated as the public sector in many developing countries withdraws from service functions, as a result of structural adjustment programmes, and other types of institution become increasingly important (Smith and Thomson, 1991).

Effective good links between farmers and all the agencies involved in providing research, extension and other agricultural services are now seen as

---

2. This assessment and the following sections are based on the results of a recent ODI study of the role of NGOs in agricultural research and extension in developing countries (Farrington, Bebbington, Lewis and Wellard, 1993).

vital. But it is recognised that such links are difficult to organise and sustain, especially for farmers in marginal, variable environments, and are strongly influenced by the particular policy and institutional context (Merrill-Sands and Kaimowitz, 1990). Many NGOs appear to have the type of organisational structure and working methods best suited to achieving such good links.

However, some problems are beginning to emerge. Where NGOs are growing in size, some of them are beginning to impose the same heavy accounting and reporting requirements on local implementing agencies as do the larger government and donor organisations. And amongst some Southern NGOs there is frustration at having to receive donor funding channelled through Northern NGOs, who they perceive as contributing little to their work except as service delivery agents (ODI, 1988). In some cases, Southern NGOs are perceived as elitist by the local communities they work with (Fowler, 1991). It is also becoming clear that the diverse demands placed on NGOs by their different clients present almost impossible challenges in terms of management skills (Brown, 1991). It is also unclear whether all NGOs have the time and resources to develop the necessary technical capabilities and links with sources of innovation (Osborn, 1990).

Thus there can be disadvantages in NGOs' organisation and method of working. They are generally small, and many have relatively unskilled staff and programmes which are chosen somewhat unsystematically and distributed unevenly among and within countries. Accordingly, their effectiveness and staying power can be limited—exacerbated by poor organisation and co-ordination. Collaboration with other NGOs and government is often weak: NGOs frequently find themselves competing for resources or setting up parallel, duplicate development efforts. For some NGOs, particularly the smaller ones, this is because they have insufficient resources to allocate to forming links or because their links are largely determined by local political, economic and historical factors. Poor links may simply be due to unfamiliarity with other agencies' operating procedures, or to differing aims and objectives.

Thus the evidence on the comparative effectiveness of NGO and official aid is so far inconclusive and inadequate. It is not clear whether the apparent advantages of NGOs are due only to their size or whether there really is an alternative NGO administrative model of development.

NGOs and Seed Supply—Modelling Interactions

One of the major purposes of this book is to establish what types of support are of most value to communities seeking to strengthen local seed systems, and thus the comparative advantages of different institutional approaches.

The reasons why individual NGOs have become involved in seed activities can be grouped into three main categories:

- **relief**: to provide relief or rehabilitation after emergencies;
- **development**: to provide access to seed along with other agricultural

inputs, often due to the perceived failure of the formal sector to reach particular communities or groups within them;

- **advocacy**: to support local communities' efforts to maintain seed themselves and, in particular, to strengthen farmers' rights to plant genetic resources.

There are two distinct rationales that NGOs use to justify their involvement in supporting local seed systems. Some maintain it is necessary in order to increase economic efficiency and growth through the use of modern varieties and better quality seed. Others justify support in terms of reducing the risk small farmers face and reducing their dependence on external agencies through supporting diversification of varieties and increased use of farmers' varieties. NGOs' seed activities can substitute for, complement or create an alternative to existing formal sector seed activities.

Some of the interactions and compatibilities between different planting material and institutional structures are outlined in Diagram 2.1. This shows the interactions between the functions that agencies aim to perform (and so the type of institutional structures that they work through) and the technologies used within small-farm communities. It also shows the influence of these interactions on community self-reliance and control of seed systems (the ability of communities to produce regular supplies of good quality seed without requiring the support or intervention of formal sector institutions). The position on this matrix of a number of NGOs currently working in seeds is shown.

Clearly, some technology systems and institutional structures are incompatible. These are represented by the shaded boxes in Diagram 2.1. For example, communities wishing to use F1 hybrid seed varieties are unlikely to be able to meet their seed needs through saving seed on-farm because of the difficulty of maintaining separate parent lines on small plots. By the same token, national seed companies are unlikely to have the capacity and motivation to collect and multiply farmers' varieties, preferring instead to concentrate on quality seed of a limited number of modern varieties. However, other combinations appear to be compatible, but have been neglected or rejected by external agencies to date. These are represented by the blank boxes in Diagram 2.1. Such combinations thus include farmer-managed seed systems working with farmers' varieties, and community seed banks working with enhanced farmers' varieties and locally adapted modern varieties. In general, agencies have been quicker to set up parallel systems (which we identify as local seed multiplication and distribution) than to work with existing systems (which we call farmer seed systems).

One important point to note is the influence of government policy: an interaction may be possible from a technical and institutional point of view, but policy directives may expressly forbid it. An example of this is the contrasting policies of different countries to modern maize varieties. Some countries, such as Zimbabwe, consider that the incremental yields possible

Diagram 2.1: Working with Seeds at Community Level: Interactions and Compatibilities

Function / Approach	Seed System / Seed Variety	Farmers' Varieties	Enhanced FVs	Locally adapted MVs	Modern varieties Self-poll/OP	F1 hybrid	Agency Contribution
RELIEF	Emergency seed distribution	*Concern, Sudan* Afghanaid, Afghanistan			*Concern, Sudan*		Northern relief agencies supplying funds, transport and personnel
	Community seed banks	*ACORD, Mali Oxfam, Sudan* NEF/Oxfam, Mali			*ACORD, Sudan*		Northern and Southern relief and development agencies providing funds and technical assistance
DEVELOPMENT	Farmer-saved seed and community seed exchange	COMPATIBLE WITH STRENGTHENING COMMUNITY CONTROL OF SEED SYSTEMS					
	Local seed multiplication and distribution	*ASS, Ethiopia MIND, Philippines* *Z-SAN, Zimbabwe*	*SCF, Gambia CESA, Ecuador FFHC, Gambia GSM, Gambia MCC, Bangladesh KHAP, Nepal AA, Nepal*	*CIC, Mozambique* *CIAT, Great Lakes*	*SCF, Gambia CESA, Ecuador AA, Gambia GSM, Gambia MCC, Bangladesh FFHC, Gambia PPS, Nepal KHAP, Nepal AA, Nepal PAI, Brazil CIAT, Colombia*		Northern and Southern development agencies providing funding, technical assistance and institution-building support
	Government/ MNC seed companies						
ADVOCACY	Seed exchange	*MASIPAG, Philippines*		*MASIPAG, Philippines*			**NATIONAL LEVEL** Southern NGOs, sometimes assisted by Northern NGOs, sharing information and technical assistance
	Training	*ASS*	*ASS*	*ASS*			**INTERNATIONAL LEVEL** Northern NGOs and NGO coalitions providing funding, technical assistance and information
	Networking	*GRAIN RAFI ASS*	*RAFI*	*RAFI*			
	Lobbying	*GRAIN RAFI*	*GRAIN RAFI*	*GRAIN RAFI*			

Source: ODI records

Notes: *Italic script = profiled in case studies*

Ordinary script = other known support for local seed systems

All acronyms listed at front of book

Combination of seed varieties and seed systems that are

▓ = incompatible

░ = usually incompatible

with strengthening community control of seed systems

This Diagram is a simplified representation of the seed activities of the agencies referred to and should be read in conjunction with the explanation and analysis given in Chapters 3-8.

from hybrid maizes are so superior that they ban the sale of open-pollinated maizes, which effectively precludes any local-level seed multiplication for maize. The opposite is true in other countries, such as Nepal, and a thriving local-level seed multiplication system is in operation for maize, whilst the import and domestic multiplication of hybrid maizes is banned.

We will be returning in Chapter 8 to the question of why some of these alternatives have been neglected so far and their possible future scope. For the moment, our aim is simply to make clear the potential interactions and compatibilities between different technology systems and institutional structures and the implications of these for strengthening community control of seed systems.

3
Seed Supply for Relief and Rehabilitation

Seeds for Relief and Rehabilitation in the Sahel of Africa

The Sahel is a vast dryland region lying between the Sahara desert to the north and the Equatorial forests to the south, stretching from the Atlantic Coast of Mauritania, Senegal and The Gambia in the west to the Red Sea and Indian Ocean in the east. Its inhabitants depend largely on cropping and raising livestock, despite the limited potential of the zone and the variability of the rains. Since the 1960s, the Sahel has suffered from a series of droughts beginning with that of 1968–73, and from declining rainfall. The people of the region face challenges of adaptation in the long run, and survival through droughts in the short run. International aid agencies, especially NGOs, have provided relief to alleviate famine quite successfully. They have also tried to assist in efforts to rehabilitate Sahelian communities in the aftermath of droughts. Of particular interest here are attempts to establish seed supplies to ensure that after droughts some planting material remains for the next crop season.

Agency for Cooperation and Research in Development (ACORD) in Timbuktu and Gao, Mali

Mali is a large (1.24 million sq.km.), landlocked country with a population of just 8.2 million (1989), more than four-fifths of them living in rural areas. ACORD, a European-based NGO, began work in the country in 1974, focusing on two of the more remote and arid regions in the north of the country, Timbuktu and Gao. Here, the 850,000 inhabitants live mainly either from pastoralism (the Tuareg and Tamacheq) or from farming and fishing (the Songhai). For both groups, the river Niger, curving through the northernmost swing of its great bend, represents a key resource. The river provides fish for the Songhai, whilst its annual floods and ebbs offer the chance to grow riverine crops of floating rice and 'bourgou', a reed which is used as fodder for the livestock of the Tuareg. Farmers also plant millet and sorghum in the drier areas of the two regions.

ACORD first supported a government programme for the re-launch of cooperatives, which had been set up during the 1960s throughout the country at official instigation. In 1976/77 ACORD helped to set up the first seed stores, with the aim of guaranteeing seed security through village-level seed stocks held additionally to household seed stores. Another objective was to free farmers from the burden of debts incurred by buying seeds from traders or landlords at planting time. ACORD provided funds to villages, where a committee was named to construct and manage the stores. Once built, the stores were stocked by seed bought at harvest time, to be loaned out at planting time and paid off in kind after harvest at a locally-determined rate, usually 150 per cent. If the harvest failed, then no interest was collected. By 1982 there were 71 stores in Timbuktu, which had in 1981 distributed 90 tons of seed to 5,800 beneficiary families; and 90 stores in Gao, which had distributed 373 tons to 16,800 families.

Droughts and poor harvests subsequently occurred in 1981, 1983, 1984, and 1985: by 1986 Timbuktu was declared a *Zone d'Urgence* with official programmes of food-for-work and food distribution.

In 1983 and 1987 ACORD evaluated its work and concluded that the seed stores met a felt need and that the programme had reached the targets set. Problems were, however, evident. Drought and harvest failure had led to the depletion of seed stocks. Seed needs were difficult to identify. Selection, treatment, and storage of seed were all either non-existent or rudimentary. Seeds were often distributed unequally according to power and political favour within the villages. The system of seed loans was hit by poor record-keeping and also led to farmers feeling indebted to the stores after harvest failures. In particular, the farmers did not feel the stores to be theirs, partly because they had not been involved in their establishment or operation.

Subsequently ACORD changed its strategy, switching from working with cooperatives towards informal village groups. By 1991 ACORD was working with 140 village groups, pastoral associations, and some cooperatives in Timbuktu and Gao, representing 150,000 people. Activities included small-scale irrigation, riverine fodder cultivation, well digging, tree planting, reconstitution of pastoral herds and the rehabilitation of seed stores.

The aims of the seed stores remained unchanged, but not the operating system. This time the seeds were to be sold to farmers for cash, even if this meant that some farmers might then only have access to seed from the village stores indirectly, via social solidarity networks. In each village a five-person management committee was given full responsibility for the store. However, the price for seed buying and selling (purchase price plus costs of seed loss, bagging, transport, etc.) is a joint decision with ACORD. The agency also provides some funds for buildings and disinfection, plus training and literacy courses to help local people administer the stores. It is hoped that these changes would make it easier for the local committees to take over fully the running of the stores. By 1991 some 84 stores had been rehabilitated under this programme.

By 1990 it was noted that, at least in Timbuktu region, stores which had problems in selling seed for cash had moved to credit sales. Others, after the good 1989 harvest, had exchanged seed with farmers, with a small margin in favour of the store, to build up stocks.

The stores have not been used as a channel for modern varieties. Links to government research bodies are weak and the modern varieties available are inappropriate, having low drought tolerance.

For the last two years the ACORD programme in northern Mali has suffered from civil turmoil in the area.

ACORD, OXFAM and Concern in Sudan

Sudan is the largest country in Africa (2.5 million sq.km.), with a population of 24.5 million, 78 per cent of whom are rural, and most dependent on agriculture for their livelihoods. Crop cultivators can be divided into two groups: large farmers with diesel-powered machinery who buy seed from parastatal agencies; and smaller scale tillers with animal draught power, who use their own saved seed or else acquire it locally through exchange or from village markets. Although there are significant areas under irrigation, the majority of farmers depend on the rains. They have suffered from a series of droughts during the last decade. In addition to droughts, the country has also faced problems presented by providing refuge for large numbers of people displaced by strife in neighbouring countries, as well as from civil war in southern Sudan. NGOs have been in the forefront of external assistance to help Sudan cope with drought and refugees and some have included seeds activities in their programmes.

ACORD in Qala-en-Nahal. At Qala-en-Nahal, in the east of the country (110 kilometres south-east of Gedarif), a refugee settlement was started in 1969, principally for those fleeing conflict in Eritrea. By the late 1980s, it comprised 11 villages housing 35,000 Eritreans and 5,000 Sudanese (many of them originally migrants or refugees from other countries and resident in the area for up to 60 years), with seven different ethnic groups represented. They were farming 16,000 hectares of black cotton soil in four-hectare family farms, producing sesame as a cash crop, and sorghum, millet, and legumes for food. They also had 80,000 head of stock. Charcoal-burning was widespread, with attendant worries about deforestation.

The Sudanese Commission for Refugees and, since 1981, ACORD, have provided services to the settlement. Apart from emergency food distribution following harvest failures, since 1985 ACORD has been active in agricultural and natural resources projects, including horticulture, forestry, livestock, credit, reduction of storage losses, control of *striga* infestation of sorghum and the introduction of new crop varieties. In the late 1980s ACORD had 18 extension staff stationed in the six largest villages of Qala-en-Nahal. By 1989, the cropping programme had tried and tested various recommendations through on-farm trials. Some involved new crops and varieties, and a strong demand for seed of these varieties developed. That year, seed from the previous year's trial sites was sold to about 260 settlers. They included legumes (tepary bean, green mung, modern and local variety cowpeas) in lots of under 1 kilo; sorghum in 6–18 kilo packs; and millet in average 6 kilo lots.

In the latter half of the 1989 season, a survey was carried out covering 216 buyers of seed from the ACORD trials and some contact farmers (55 per cent of them women). Information was gathered about seed sources, practices, and demand. Analysis of the results suggested that local knowledge about seed selection and preservation was limited, partly because some settlers came from pastoralist backgrounds, or because they had spent time labouring on others' farms and had not been concerned with seed. The survey found that about a quarter of farmers were using seed from other farms, either from individuals with a reputation for good seed quality, or from markets where dubious, nondescript 'seed' was on offer. It also confirmed that there was strong demand for the seed from ACORD trials.

ACORD therefore decided to establish village seed banks. These would provide local outlets for improved variety seeds, improve the supply of reasonably priced quality seeds at planting time, and encourage local enterprise. It was planned to train farmers in seed technology. The villages were to set up seed bank committees, to which ACORD would give up to US$240 to match locally-collected funds, seeds from the trials at subsidised rates, technical assistance, and, at cost price, storage equipment, Aldrex insecticide dressing and stationery. By early 1990, four committees had been formed, and they had bought most of ACORD's seed as well as acquiring extra seed from the National Seed Administration, a Canadian farming project, a research station, and from local farmers.

Unfortunately the 1990 harvest failed badly, and the seed banks lapsed: indeed ACORD had to send funds to help buy seeds for the next season. Moreover, many of the Eritreans have left Qala-en-Nahal to return home following the change of government in Ethiopia.

OXFAM in Kebkabiya. In western Sudan in the province of North Darfur, OXFAM, a UK-based NGO, funds the Kebkabiya Smallholders' Project (KSHP). This serves 16 groups of villages ('centres') with a population of 10,000 families, aiming to improve local livelihoods in an area highly susceptible to drought and declining rainfall. The project is managed jointly by a committee of volunteers from the centres and by OXFAM's hired manager, supported by a team comprising a veterinary officer, an agricultural extensionist, a pest control officer and women's co-ordinators.

Seed banks were first built in 1985/86 in each of the centres: these were stone stores constructed by communal labour using local materials and cement. They were stocked with enough millet seed to provide each family with 5–10 kilos of seed, sufficient to plant 1 to 2 ha of millet. Seed taken from the banks was repaid after harvest with an in-kind interest of 1.8 kilos. For 1986/87 and 1987/88 this worked; indeed in some centres it was so successful that returned millet seed exceeded storage space and was sold off. However, the 1988/89 harvest was poor, the first of a series of bad cropping seasons, and by 1990 the stores had been emptied.

In late 1990 the KSHP decided to restock the stores and undertake a crisis distribution of millet seed. The management planned to buy millet seed for 10,500 families locally, using a supplementary grant of more than US$80,000 from OXFAM. The seed would be given to families either on credit or for cash. Seed distribution was to be administered by the 16 centres who would control the credit: proceeds of the sales were to be banked separately to create a fund to cope with future droughts.

In the event, seed was bought in Nyala in South Darfur (four hours drive away) since local stocks were inadequate, and grain prices in Kebkabiya were soaring. The seed cost more than expected: budgets had been drawn up at post-harvest prices but owing to a delay in obtaining funds the seed was eventually bought four months later when millet prices had doubled. Consequently, less seed could be purchased than had been planned (down from 79 tonnes to 42 tonnes); some groundnut and cowpea seed was also added to the millet. This was transported back to Kebkabiya, sent out to the centres and exchanged for seeds of local millet varieties. Farmers prefer their own varieties of millet, and are accustomed to substituting local seed for 'foreign' seed and eating the latter. They are also used to exchanging seed with neighbouring farmers.

The exchanged local variety seed was made available to households in time for planting at the rate of 4–9 kilos of seed each. In 14 out of 16 centres, all the seed available was taken up. In the other two the policy of charging to cover the cost of seed purchase plus wastage pushed prices above those in the local market, and seed remained unused and was reallocated among the other centres. Three-quarters of the seed was sold for cash, the rest on credit.

The seeds were planted, thanks to having been distributed in May 1991, just before the start of the rains, but again the rains failed and the harvest was poor—satisfying perhaps 20–40 per cent of food needs. At the end of the season most of the credit had not been repaid.

Later during the season (July–August) another NGO, Save the Children UK, gave out seed free in Darfur as emergency assistance. Since this seed was neither local, nor in most cases made available in time for planting, most of it was eaten.

The KSHP distribution was successful, and demonstrated the ability of the committee to manage such a programme. There remain, however, the problems of what to do about the credit which is unlikely to be repaid, and of how to administer revolving funds when inflation is undermining the value of local currency.

Concern in Kosti. During the 1985 food crisis in Sudan, an Irish NGO called Concern began distributing World Food Programme food in Kosti Province (then known as South White Nile Province) in central Sudan. It also delivered seeds—300 tons of sorghum, 280 tons groundnut and 18 tons of sesame—to ameliorate the shortage of seed from the preceding bad harvests. Subsequently, Concern followed up its relief activities with a development programme, including education, nutrition, primary health care, reforestation and agricultural extension (crop husbandry trials in 13 villages, propagation and provision of early-maturing sorghum, agroforestry and schools plots).

The rains, however, were poor in 1989 and 1990 leading to small harvests, depleted seed reserves and high prices for the little seed available on the market. By the second half of 1990, Concern was planning an emergency relief programme for the province, consisting of food distribution, supplementary feeding, food-for-work and seed supply. Appeals for support were made against the background of Sudan's increasing isolation from American and European donors, exacerbated by the tensions of the Gulf Crisis. Nevertheless, humanitarian criteria eventually prevailed and assistance was secured from USAID and the EEC.

Concern planned to provide about 30 per cent of the total seed requirement for three-quarters of the rural population of Kosti Province, with an initial target of 642 tons of seed. Thanks to having agricultural programme staff, Concern was able to assess the need for seed and to specify the mix of seeds appropriate for different villages with respect to three soil types (clay, sandy and clay-sand) and two rainfall levels (for the north and south of the province). Seed was procured in April and May 1991 from two sources. Half came from the Agricultural Bank of Sudan (ABS), secured through Department of Agriculture requests, out of stocks held at Gedarif and Sennar. This was sampled and tested for germination (90 per cent) and purity (85 per cent) at the Sennar labs of the National Seed Administration. ABS seed was bought at prices 25–50 per cent below open market levels. The other half came from local merchants within the province, some of which was sampled and tested by Concern staff.

Concern recognised that local seed varieties, usually early-maturing and with good drought resistance, were preferred to the ABS stocks which were of medium duration and drought tolerance. However, local varieties cost 35 to 45 per cent more and merchants had only limited quantities.

In the end, 460 tons of sorghum, 279 tons of millet, and 131 tons of sesame seed—870 tons in all—were obtained, considerably more than was initially planned. This reflected an increase in the number of villages included in the programme. Not only did the amount of seed bought increase, but also prices proved to be higher than expected. The total cost of the seed and its distribution was US$1.36 million.

The programme was coordinated with the Sudanese government Relief and Rehabilitation Commission and the Department of Agriculture, and the regional activities were directed by the provincial Relief Committee. Technical inputs were obtained from both the local Department of Agriculture offices and the National Seed Administration.

The seed was neither processed nor treated, partly due to lack of time, partly because farmers were used to imperfect seed, and partly because of the dangers of applying chemical coatings to seed which might be eaten. Most was delivered to the 660 Village Relief Committees (VRC) although some went to some semi-urban farmers in the three towns of the province, and to two camps for the displaced. The VRCs then distributed the seed, apparently equitably, among the local population of some 67,500 farm households. Delivery was made in June, a few weeks after a double ration of food aid had been handed out, thereby reducing the chances of the seed being eaten and, in the event, three weeks before the (delayed) rains began. Seed packs were around 10–15 kilos per household, enough to plant 2–3 hectares, about 20 to 25 per cent of normal requirements.

Unfortunately, the rains, having begun late, ended early in August and the harvest was meagre: only those farmers who had planted early-maturing varieties harvested anything. Farmers had wanted these varieties, but they had been in short supply. Indeed, farmers expressed a marked preference for local seed, even when that from afar was of the same variety.

Concern later proposed that the seed handed out should be paid back in kind after the harvest to the VRCs, with the idea of forming community seed banks. The farmers were not all aware of this and the widespread harvest failure prevented any significant return. However, about 5 per cent of farmers in 18 villages achieved sufficient harvest to supply sorghum seed, in return for food aid wheat. Concern arranged for this seed to be stored within the communities and it was distributed by the VRCs for the 1992 season. The communities themselves collected money to buy anti-termite seed dressing. Concern is now waiting to see whether the seed will be re-paid after harvest.

At the same time, Concern arranged a big distribution of 450 tonnes of sorghum seed that it had purchased from outside the area, for a second year running. This time, however, it was able to purchase local early-maturing varieties from mechanised farmers in the South of the province and it targeted the seed, giving different quantities to three different categories of community.

Rehabilitating Local Seed Supply in Mozambique

Mozambique is a large country (800,000 sq.km.) on the East African coast, with a scattered population of 16 million, half of whom are thought to have fled from their original birthplace to escape the on-going war between the government and RENAMO rebels.

With average per capita incomes of only US$80, the economy, based on agriculture and transport to the country's land-locked neighbours, is facing a very difficult period of structural adjustment and political reform, exacerbated by the war and the lack of development prior to independence in 1975. There is a high degree of dependence on international donors, including NGOs, in many sectors.

Only about 4 per cent of the total land area is cultivated and the highly variable rainfall constrains the farming system, which is dominated by maize, sorghum and beans. Wheat is also grown in some areas, as are sugar and other cash crops on large state farms. Most of these have now been divided up and the emphasis is on small, low-input family farms.

Recent estimates [DANAGRO, 1988] put the amount of certified seed required for the whole country at some 19,600 tonnes per year, with just over half needed for the small family farm sector, mainly maize, sorghum, rice and beans. Current certified seed usage, mainly rice, is about 14,000 tonnes per year, of which nearly 8,000 tonnes is imported, up to two thirds as part of donors' emergency programmes.

The organised seed sector has had substantial support since the late 1970s as part of the Mozambique Nordic Agriculture Programme (MONAP), with the aim of ensuring all sectors use certified seed of modern varieties, particularly maize, rice, sorghum, wheat, beans, groundnuts and sunflower. The intention, except for wheat, is to provide all of these from domestic production, to end dependence on imports which currently cost US$4–5 million annually and are prone to quality problems.

A national seed company, Empresa Nacional de Sementes (ENS), was formed in 1981—succeeded in 1988 by Sementes de Mozambique Lda (SEMOC)—to provide processing plants in each province supported by local seed production capacity. SEMOC is now operating from four sites in the south, north and west, concentrating on maize, rice and groundnuts, with potato seed production being developed. The government is encouraging local associations as the most appropriate institutional structure for seed production and distribution, with SEMOC undertaking collaborative ventures.

Substantial progress has been made in setting up an organised seed sector and many of the necessary supporting services, for example variety development, are now well-focused on the needs of small, low-input family farms. However, the capacity to reach small farmers with new seed remains low. This is partly the result of Mozambique's historical under-development but it has been exacerbated by the war, which prevents on-farm agricultural research trials and has destroyed newly-installed seed processing capacity in the provinces. The war also limits farmers' ability to make effective use of new seed and therefore their demand for it. It has also resulted in a parallel seed distribution network operated by donors' emergency programmes, which often end up competing with each other and with the national programme: unknown quantities and qualities of seed are distributed free by a large number of organisations.

National policy is therefore now promoting the distribution of seed from sources nearest the end user and the control of emergency seed distribution, to ensure it fits in with normal market systems as far as possible.

It is in this context that Centro Internazionale Crocevia (CIC) has been working in Niassa Province in north-west Mozambique since 1981, in various development projects in agriculture, education and water supply. CIC is an Italian non-profit association set up in 1958, linked to the Regional Centres for Technical Education in Agriculture in Italy. It aims to assist the autonomous development of communities in developing countries through supporting local control of genetic, cultural, energy and environmental resources. It works with Southern partners such as farmers' associations, NGOs, teachers and some governments and it has US$3 million committed to genetic resources development in Burkina Faso, Nicaragua and The Philippines as well as in Mozambique.

Niassa was once Mozambique's best area for maize, beans and potatoes but its economic and social infrastructure was largely destroyed by RENAMO activity between 1984 and 1986. Many lives were lost, food supplies threatened and its rail links with the rest of the country were cut. The area is atypical in that it is high (1,300 metres) and has a temperate climate.

The project CIC supports, the Niassa Seed Production Board (GPSN), aims to establish a profitable seed organisation independent of central government, to support long-term local control of the seed chain. The main activity is seed multiplication but it is also moving into research: many varieties and minor crops which were important locally before the introduction of maize are disappearing and the project is trying to protect these resources by encouraging their economic use, not simply by conservation. It is planned that the major shareholders in GPSN will be SEMOC, the Niassa Department of Agriculture and the local private sector, including GPSN workers. The project has funding from CIC, from the local counterpart (the Ministry of Agriculture), and US$2 million over seven years from the Italian Ministry of Foreign Affairs and the EC. In addition, two local banks have provided working capital for buying seed from contract growers.

CIC, which has provided technical assistance and capital items, is the only NGO involved in the project. Its links with government are regulated by a co-operation agreement. The project involves close links with government and the other seed organisations. GPSN, which is under the control of the local government, has done most of the planning and co-ordination work although the intention is for it to be a 'service' organisation for a network of seed multipliers.

The aim is for the project to form a component of the national seed production plan, and it has an important role for crops, such as wheat and some vegetables, which are not grown widely elsewhere in Mozambique. But there were initial problems with having the project included in the national seed production plan. Even now donors and other seed distributors make only limited use of the project's output as they negotiate the areas to receive seed and the quantities to be distributed centrally with a national Commission unrelated to the Department of Agriculture. This has meant that GPSN has faced obstacles in selling seed outside Niassa Province and that there has been some reluctance to allow it to take over responsibility for supplying Niassa.

Basic seed production, experimental testing of local varieties and production techniques and seed processing are carried out on a 15 hectare site 15 kms from the provincial capital, which was donated by SEMOC in 1987 after bandit activity caused it to be abandoned as one of the company's sites. Certified seed is multiplied for GPSN under contract, mainly by local state farms but also by co-operatives, private enterprises and farmers, who also distribute seed. The average number of growers each year is 10. Much production is mechanised; government seed inspectors carry out full field inspections; and seed is cleaned and treated at the processing plant, which employs specialist seed staff trained within Mozambique and at the national seed companies of Zimbabwe, Zambia and Kenya.

Farmer participation is recognised as important but it is made difficult by the large size of the province, the poor transport and communications network and the tradition of self-sufficiency forced on local farmers by the war. GPSN also encourages worker participation via company shares and by encouraging workers to use the project farm to grow their own choice of varieties, to compare with colleagues and the organised trials.

Seed production started in 1988 and has involved a number of local and modern varieties of maize, beans, soyabeans, sunflower, wheat and sorghum (see Table 3.1). There has been a reduction in the number of varieties grown over time but an increase in the range of crops. The area has remained constant, at around 70–75 ha, but clean seed production has increased dramatically from 40 to 180 tonnes.

The continued dominance of maize in the production programme is an anachronism dating from the time when the major market for GPSN seed was

Table 3.1:		GPSN seed production 1989–91					
Crop		Area (ha)			Production (tonnes)		
	1989	1990	1991		1989	1990	1991
Maize	53	52	–		37	129	150
Beans	15	10	–		–	5	–
Soyabean	2	3	–		0.8	3	3
Sunflower	0.5	3	–		–	2	1
Wheat	1	5	–		1.4	15	30
Sorghum	–	–	1		–	2	–

Source: Gaifami, 1991b
Varieties: Maize: Obregon, Ferke, Manica
        Beans: Manteiga, Dotor, various
        Soyabean: Oribi, Hardee
        Sunflower: Peredovic, Elias
        Wheat: Kenya Nyati, Loerie
        Sorghum: Serena, local
Notes: Production of clean seed only; part of wheat area de-classified in 1989;
        – = not available.

the state farms. Wheat seed in particular is needed in Niassa: because GPSN seed costs only $US0.25/kg to produce whereas imported wheat seed costs US$1/kg; and wheat is not grown elsewhere in Mozambique, so there is no priority on wheat breeding in the national agriculture research programme.

The varieties produced largely meet farmers' needs in terms of being low-input, but the project has test plots of many other varieties and crops, including green manure, vegetables, potato, bambarra nuts, fodder, pigeon pea, groundnuts, sesame, cassava, chick pea, coriander, hyacinth bean, other local beans and local varieties of maize, sorghum and groundnut.

There is a limit on total production of 250 tonnes per year imposed by the absorptive capacity of Niassa region: the processing plant itself has a potential throughput much greater than this. The plan is to maintain maize and soyabean seed production levels but to increase production of sunflower and wheat and beans. Production of sorghum seed is popular with growers but no variety is currently available that meets the needs of local buyers.

The target group for seed distribution is the family farming sector, because this is of greatest economic and social importance in Mozambique. But the families to which the project distributes seed are not typical of rural conditions in Niassa because they are all within 50 km of the provincial capital. The lack of transport infrastructure and commercial activity outside these areas prohibits wider distribution. The aim is to support what remains of the existing commercial distribution network, although this has been severely damaged both by war and by emergency programmes, rather than to set up parallel distribution systems. Therefore there is no free distribution and there is no direct distribution by GPSN, even though this would be more profitable. It is estimated that around 200 families have obtained seed through this system. However, many thousands of families have received GPSN seed as part of emergency distributions by other donors.

Overall, seed sales have been very disappointing, with stocks left over in 1988 equivalent to 70 per cent of production. Poor sales were caused by the lack of commercial trade in Niassa Province, by high retail seed prices compared to average farm incomes (although they are some 30 per cent lower than SEMOC's prices) and by competing free seed distribution by other donor agencies. Sales are also limited by the lack of extension services in the province, which mean a lack of awareness of the benefits of new seed. Part of the problem at the moment is the danger, due to the war, of travelling in the rural areas.

The project aims to be sustainable in the future without external intervention. Seed production can provide growers with a good income and GPSN gives them equipment and crop protection chemicals on a loan basis. If there is sufficient local interest in new seed, the project should be able to operate without subsidies or other support, as seed production costs are lower and yields are higher than the national average. However, the project needs to sell 1,000 tonnes per year at current prices in order to break even; in 1988/89 it had a US$12,800 deficit. In addition, although foreign exchange needs are small and in the long run can be met from earnings from seed sales to donor agencies, short-term support is needed for technical assistance salaries and imported equipment. Interest payments on loans from SEMOC and local banks, taken out when funds expected from the Italian Ministry of Foreign Affairs arrived late, have become very large, so local staff have had to go back onto the government payroll.

4
NGOs in Local Seed Supply in Latin America and Asia

Potato Seed Multiplication in the Ecuadorian Andes

Ecuador, with a population of 10.3 million, 45 per cent of them living in rural areas, is a country of contrasts. Geographically, it varies from the coastal plain to the Andean mountains, and the eastern forested lowlands of Amazonia; socially, there are marked disparities between the better-off and the poor. Poverty persists despite Ecuador's status as a middle-income developing country and its growing economy, which produced a GDP expanding at 3 per cent per year from 1965 to 1989.

Social differences are reflected in the dual nature of the country's agriculture. On the best soils there are large farms with access to inputs, favoured by state policies in technology development, which produce a wide variety of exports: beef cattle, cocoa, coffee, soyabeans, fruits, cut flowers, honey, and bananas. On poorer land and predominantly in the Andes, smallholders grow food crops such as potatoes, barley, beans and vegetables using traditional techniques. They find it hard to obtain inputs and have had little attention from governmental research and extension systems. Moreover, they face a twin challenge from an increasing rural population and consequent division of the land into ever-smaller farms (*minifundia*); and from degradation of soils as the intensification of cropping with shortened fallows, and declining numbers of stock held per farm and less manuring, lead to an 'organic matter crisis'. In response, smallholders have bought in chemical fertiliser, domestic fuel and housing materials to replace items once produced within the local farming system. Consequently, Andean peasants have had to sell more to survive—a burden aggravated by the terms of trade moving against their products during the 1980s, while the state reduced still further the limited services offered to Andean smallholders in an effort to trim government budget deficits.

Throughout the highlands, farmers have formed their own associations and unions to obtain services and rights. Campaigns for land reform in the 1960s and 1970s were a potent reason for organisation; but others include the provision of electricity, the price of public transport, the defence of religious freedoms, and co-operativism. Subsequently, these organisations have moved on to pursue other objectives, including the improvement of local agriculture, for which federations of farmers' groups have hired professional agronomists to help them. The federations, in turn, often collaborate with and receive assistance from one or other of Ecuador's many independent NGOs or 'grassroots support organisations'. These provide funds, technical assistance, and linkages to national resources, including government organisations such as the national agricultural research institute (INIAP), as well as representing peasant interests in national debates.

CESA, the Ecuadorian centre for agricultural services, is one of the largest of the national NGOs concerned with smallholder farming. Founded in 1967, its activities include agriculture (production support and technical assistance, provided by a professional staff of 23), infrastructure development, conservation of natural resources, reforestation, training, and promotion of women's issues. It operates in 10 areas of the country, mostly in the Andes, and works with around 10,000 families who belong to 120 producer groups. CESA's aims include strengthening farmers' organisations, both to manage their own collective affairs as well as to enable them to negotiate with the state and its institutions about services and policy. These aims are reflected in the way that CESA works collaboratively with farmers' groups, responding only to those needs and priorities signalled by them, and adopting participatory methods for research and extension. In 1979 the agriculture programme began an experimentation and demonstration project, funded by German Agro-Action. It is within this programme that CESA promotes seed supply.

Seed in Ecuador is officially supplied by INIAP as basic and registered seed to commercial companies and to farmers. Little of this, however, reaches smallholders since stringent quality standards, plus the costs of distributing seed in highland terrain in small packets, raise the prices of commercial seed above what smallholders are prepared to pay. Moreover, modern varieties released through the Ministry of Agriculture as a public service go to only a few of the better-off peasants able to grow seed crops to INIAP standards, and even this is not widely distributed. Consequently, although 70 per cent of the main Andean food crop, potatoes, is produced by smallholders, fewer than 3 per cent of them use registered or certified seed. Yet better potato seed is needed. Local varieties have deteriorated in recent decades, partly because of fungal and other diseases infesting stored potato seed.

CESA helps Andean farmer groups multiply up potato seed, and some barley seed (as well as maize and rice in other parts of the country). Activity begins with discussions between CESA and a farmer group federation. Agreement typically results in a community plot of up to two hectares being assigned for seed production, whereupon for the first year CESA provides seed and agro-chemicals whilst the federation supplies the labour. At the end of the season, CESA recoups its investment in harvested seed, the rest of the crop being both sold in the market and divided amongst the federation's membership with each member typically getting 45 to 90 kilos of seed for planting, depending on the amount of labour contributed. Seed is sold to the members at market price less five per cent. CESA also trains local farmers to run the multiplication plot in the future.

Planting material is either modern varieties, potentially higher-yielding but risky, or else selected from local stocks (favoured for their hardiness and taste), taking particular care to ensure that it is free from pests and diseases—in which task CESA gets help from the university. In either case, CESA tries to produce 'artisan' quality seed, that produces better planting material for local farmers at affordable cost, but below the strict standards of official certification.

Seed is kept in simple, cheap sprouting stores where sprouting and pests and diseases are controlled.

By late 1991, CESA's efforts had been restricted to a few years' experience, with five hectares under multiplication in three pilot areas. Typical production costs are given in Table 4.1. Seed yields in Cotopaxi Province were lower than in Canal, at 5.45 and 7.04 tonnes per hectare. In these cases, the amounts retained to continue communal multiplication, distributed to members, and sent to market were 1.35, 2.79, and 1.31 tonnes; and 0.9, 2.25, and 3.89 tonnes, respectively.

CESA's experience is not isolated: many of the peasant federations are engaged in distributing seed amongst their members, including above all potato, but also including onion, garlic, barley, beans, other tubers, and garden vegetables.

CESA has formed links upstream in seed multiplication. In September 1991 it signed an agreement for preferential access to foundation seed from INIAP—formerly CESA had faced stiff competition from private seed companies that tried to acquire all INIAP's foundation seed. The International Potato Centre (CIP) is trying to establish a national committee for potato seed production, bringing together government, private companies and the NGOs. Initial meetings have been held, but there is tension with the private companies that wish to corner the market in foundation seed.

Table 4.1:	CESA potato seed production costs, 1991	
Item		*US $/ha*
Ploughing, ridging		85
Seed	(1,260 kg @ 0.22)	280
Fertiliser	(990 kg @ 0.24)	237
Other agro-chemicals		96
Labour	(132 days @ 1.15)	152
Transport		8
Total		858
Seed yield (kgs)		8,100
Seed cost per kg		**0.11**

Notes: Variety = Bolona (local); production area = Canal.

In the long run, CESA hopes to empower the local communities, through their associations, to select and multiply seed independently. NGOs have had some success in demonstrating an alternative model to the state institutions, for INIAP has itself begun two 'artisan' seed production projects to multiply seed on poor farmers' fields.

Vegetable and Soyabean Seed Multiplication in Bangladesh

Bangladesh is a populous densely settled country (with 113 million inhabitants within its 144,000 sq.km.), and it is one of the world's poorest (per capita GDP of US$180 in 1989). About 84 per cent of Bangladeshis live in the rural areas, where farming concentrates on growing food grains, with fully 80 per cent of the crop land sown to rice. Agricultural output has increased during the last two decades thanks to the adoption of Green Revolution technology, largely for the winter crops of wheat and rice. Grain output has risen at 3 per cent per year. The gains have been skewed, however, since access to land in the Bangladesh countryside is inequitable, and population increases have eroded the potential increase in rural wages. Hence most rural Bangladeshis still live in poverty.

The Mennonite Central Committee (MCC), an NGO based in the USA and working in more than 50 countries worldwide, has been working in Bangladesh since 1970, when it provided relief assistance after cyclone damage. Subsequently it has evolved a programme which aims to improve the incomes, productivity, nutrition and living standards of rural women, marginal farmers and landless labourers. These aims are pursued in a three-part programme: job creation; health, education and social services; and agriculture. The programme was staffed in 1991 by 200 local personnel and 30 expatriate volunteers, and with a budget of around US$1 million a year (excluding the value of wheat in food-for-work projects and expatriate salaries).

Agriculture's share of the resources is US$400,000 and a staff of 135 local personnel and 18 expatriates. The agriculture programme began in 1972, focusing on the irrigated production of crops nutritionally complementary to rice, especially vegetables, during the dry winter season. It also concentrated on the newly-formed alluvial lands in coastal areas, vulnerable to flooding and cyclones and a refuge for the landless poor. Since 1972 activities have developed both technically, in adopting systems approaches to adaptive research and extension with increasing farmer participation, and socially, in targeting women, marginal farmers and landless labourers. By 1991 the agriculture programme consisted of seven areas: extension, farming systems research, partnership in agricultural research and extension (PARE), soyabeans, rural savings, homesite/women's activities and agricultural training, plus administrative support.

MCC works mainly with two kinds of seeds in Bangladesh—soyabeans and vegetables—although it also contracts farmers to multiply improved rice varieties not readily available from the government.

Horticultural seed

MCC works with 10 types of winter and 13 types of summer vegetables, principally aubergine, cauliflower, cabbage, carrot, kohl rabi, radish, and tomato. The winter vegetable seeds are generally imported, hybrid, certified seeds; the summer vegetables are mainly local varieties, albeit selected and improved through research. The extension sub-programme, staffed by a leader, five programme officers, and 22 extensionists, takes the lead in horticultural seed activities. Most of the seed, both imported and local, is purchased from wholesalers. It is then tested by MCC for germination, and repackaged in small packets (average of 11 grams per crop) for sale to target farmers, via MCC extension agents. They pay cash (although some extension workers extend credit informally on trust) at prices which cover the costs of seed acquisition and most of MCC's direct costs, but not overheads. The extension sub-programme serves about 2,000 farmers in this way. It also procures, tests, and packs seed for the rural savings and homesite/women's projects which reach another 500 targeted farmers. It also provides some seeds for the PARE venture which then sells the seeds on to partner NGOs.

The extension programme contracts growers to multiply some of the local, non-hybrid vegetable varieties (especially kangkong, borboti, Indian spinach, bitter gourd, cucumber, and okra), for which good quality seed is difficult to procure. Contracts are specified by price and area sown. MCC provides neither inputs nor a subsidy, but offers technical advice. Some farmers have been able to start seed multiplication businesses, supplying directly to local markets.

In 1988/89 24 kilos of winter vegetable seed were sold, and 271 kilos of summer vegetable seed. Seed multiplication yielded 338 kilos against a target of 277 kilos.

In addition, MCC has provided seed packets for emergency relief: in September 1988, 625,000 packets of vegetable seeds were packed for the victims of floods. In 1990, 60,000 packs of mixed seed of eight vegetables were prepared for cyclone victims.

Soyabean seed

MCC's involvement with soyabeans is in marked contrast to that with vegetables. It initially became interested in soya as an alternative crop because of its excellent nutritional qualities, and so in 1975 it joined in a collaborative project led by the government's Bangladesh Agriculture Research Council (BARC). The project, however, was closed down in 1981 leaving MCC, in conjunction with the Bangladesh Agricultural University (BAU), to carry the responsibility for developing the crop. MCC then undertook a large, integrated effort to promote soyabeans, including varietal research, seed multiplication, extension, market promotion and soya-food development. It succeeded in introducing an Indian variety, Pb-1 (officially registered in Bangladesh in 1991 under the name Shohag), with better seed storage capacity and quality, and in stimulating demand from the snack-food manufacturers. As a result, soyabean planting increased from 110 hectares in 1987 to 480 hectares in 1989. In that year the crop was added to the official Crop Diversification Programme (CDP), and a five-year action plan prepared. MCC is now collaborating with the CDP. The main joint effort in seed work is the establishment of a cold warehouse to provide storage from one winter season to the next in MCC's main winter production area. The other major collaboration with CDP is assisting soyabean food market development, a critical element in making the crop sustainable. NGOs working in other parts of the country have also shown an interest in promoting soyabean cultivation, which might allow MCC to withdraw from its leading role. Most of the soyabean producers are relatively prosperous farmers, and MCC would rather deal with soyabeans in the context of poverty alleviation.

Initially MCC contracted growers to grow seed during the summer rainy season in Chuadanga, one of the drier districts of Bangladesh, ready for winter sowing in the south-eastern districts of Comilla, Feni, Lakshmipur and Noakhali where MCC's other agricultural projects are concentrated. Subsequently, MCC's primary seed multiplication unit has moved to Tangail, where both dry winter and rainy summer production can take place. From 1990, incentives were paid to growers to improve seed crop yields. MCC is also making soya seed available to farmers through private dealers in Chuadanga, subsequently buying back the seed crop from the dealers. Planting seed for multiplication is obtained from Noakhali. In the 1988/89 season 62 hectares were under soyabean seed production; the following year 121 hectares were targeted. MCC takes most of the collected seed to the south-east for sale, now entirely via dealers, to more than 6,500 farmers.

Vegetable seed is stored at low humidity, tested before sale and packaged. The quality is high. Soyabean seed storage, however, presents difficulties. Currently most of the soyabean seed from the drier part of Bangladesh is sold for winter planting within two or three months of the summer harvest. Trying to hold seed stocks longer, especially through the summer rains, has been difficult; MCC has experimented with a variety of drying and packing techniques to improve storage, but without finding a convenient solution.

Costs of MCC seed activities are as follows. On about US$13,000's worth of vegetable seeds, direct costs in testing, handling, packing, and transport come to US$1,300, only a part of which is charged to farmers. A further US$9,200 go on staff time and management, indirect costs which are not charged to farmers. For soyabeans, about 50 tonnes of seed is bought from growers at US$300 per tonne, giving a value of US$15,000. Subsidies include: US$20 per tonne to growers for incentives to control pests and replacement of seeds lost to rains; and US$40 per tonne to buyers of soya seed, as MCC absorbs at least part of transport, storage, processing and packing costs. Total direct costs of the soyabean seed programme are thus estimated at US$3,000. On top of these come the indirect costs of staff, management, and marketing costs at US$18,000, to which might be added soyabean research at US$10,000, and expatriate costs of US$6,000 per year. The soyabean programme is a fully integrated effort to promote the adoption of the crop, so these indirect costs support this effort as well as the seeds activity. Prices charged for seed are comparable or lower than market prices for similar seed.

However, it is planned that the soyabean seed activity will become more commercially oriented, in order to encourage future private sector involvement. Accordingly, direct subsidies to growers for pesticides and source seed are being ended and the sale price of seed is to be set at a higher level, reflecting all direct seed activity costs.

MCC is involved in all aspects of seed supply: varietal research, multiplication, processing, storage, testing and quality control, and distribution—and these are further linked to MCC's agriculture programme. Staff estimate demand for seed and the amount to be purchased or multiplied with minimal farmer involvement. Plans are annual for horticultural seeds but a five-year plan has been drawn up for soyabeans. Variations in the weather mean that plans need frequent revision. Records are kept, and activities monitored. Vegetable seeds are central to the extension sub-programme, so that problems are reported speedily. The soyabean programme is less strictly monitored. Systematic evaluation of the seed programmes is not yet carried out.

The MCC wants to avoid farmers becoming dependent on the seed supply it operates. It only sells seed to target group members, and not to the general public. Cash payments for seed are the policy. The subsidies on direct costs of high-cost, imported hybrid vegetable seeds are expected to be removed during the next few years. The long term aims are twofold. On the one hand, farmers will be encouraged to save, process, and store more of their own seed. On the other hand, it is hoped that farmers' groups will buy in bulk from local dealers.

MCC has found it hard to persuade the government's National Seed Board (NSB) to release officially varieties which have succeeded, the main example being the *Pb-1* soyabeans which were not released until 1991.

Overall MCC has been successful in establishing seed supplies both for kitchen gardens and for the more commercial production of soya. Farmers are keen to obtain the seed, especially of soyabeans for which the MCC is virtually the only source in the country. Technical problems and shortage of funds hamper the soyabean activities, but the latter may be eased now that soya has been added to the government programme which has official donor backing. Otherwise, the challenges for MCC are to institutionalise its efforts with local farmers and to stimulate private dealers to increase the quantity and quality of seed that they supply. In recent years, some progress has been made towards this end and MCC no longer deals directly with farmers in buying and selling seeds and instead deals with local traders.

5
Local Seed Supply as Government Policy in Africa and Asia

NGOs and Seeds in The Gambia

The Gambia is a small country (11,300 sq.km.) in West Africa, the large majority of whose 900,000 or so inhabitants depend on agriculture for their livelihood. Most farming is carried out by small farmers in rainfed conditions with simple technology. They grow rice, the main food crop, in the swampy lowlands, and millet, sorghum, maize, and groundnuts—this last being the main cash crop and export from the country—on the higher ground. There is a marked sexual division of labour in farming, women being concerned with lowland rice and men with the upland crops. During the 1980s, agricultural output grew sluggishly, while it became apparent that, as in some other Sahelian countries, rainfall was declining and becoming less reliable and the rainy season was shortening, resulting in more frequent harvest failures.

Most of the seed used in Gambian cropping systems is saved by farmers from their harvests, with small amounts being obtained through informal exchanges and trading. While this supply works adequately much of the time, there are difficulties. Farmers have trouble in storing their seed, a problem only partly relieved by a government programme which built collective seed stores in more than 570 villages. Moreover, after poor harvests, both the supply and quality of farm-saved seed decline, and farmers are forced to search for whatever is available in the markets. Changes in local growing conditions—the decline in rainfall is a recent and dramatic example—also create a demand for new varieties from time to time.

Formal seed supply in The Gambia is little developed. Before 1985, the main suppliers were the parastatal monopoly The Gambia Produce and Marketing Board (GPMB) operating in close coordination with The Gambia Cooperative Union (GCU) for groundnut seed, and the Department of Agricultural Research (DAR) working through the Seed Technology Unit (STU) at Sapu station, which multiplied up groundnut, maize, and rice seed. Both of these government agencies were costly and depended on subsidies. Private sector retailing of seeds was limited to a small trade in imported vegetable seeds.

Following the start of an Economic Recovery Programme in 1985, subsidies were cut and the GPMB-GCU groundnut seed supply came to an end. In 1986 a new seed policy was announced, abandoning public sector production in favour of the private sector and NGOs. This also meant a change from centralised, large-scale multiplication to decentralised, small-scale operations.

Government's role, largely implemented by the STU, was to be limited to providing foundation seed to people or organisations capable of multiplying up seed and offering seed testing, inspection, processing and advisory services.

Subsequently, two sets of actors have taken the lead in formal seed multiplication and distribution: NGOs; and government projects funded largely by donors. In the late 1980s, NGOs in The Gambia were encouraged to provide all kinds of services which the hard-pressed government was finding ever more difficult to fund, and their work and numbers burgeoned throughout the country. In agriculture, there are at least eight sizeable NGO programmes, of which four have significant seed activities: ActionAid–The Gambia (AATG), Freedom from Hunger Campaign (FFHC), the Good Seed Mission (GSM), and the US Save the Children Federation (SCF). Government activities include the STU which contracts a few private growers in the Sapu area to multiply seed; an FAO-funded fertilizer project; plus—of lesser importance—a few rice development schemes and a regional development programme.

The AATG, FFHC, and SCF seed programmes originally grew out of emergency schemes to replenish farmers' seed stocks in drought years in the mid-1980s, but soon progressed to the regular multiplication and distribution of crop seed. All three programmes obtain foundation, sometimes registered, seed from the STU and distribute it to local growers, dispersed in different villages, for multiplication on small plots, usually of less than one hectare. The NGOs were already working with village groups, to whom they now channel the seed, usually with fertiliser, on credit. Initially they tended to favour multiplication on community fields, but as it became clear that these were given a lower priority than individually managed plots, they switched to private multiplication, entrusting the seed to the better farmers. In one case, FFHC subsequently switched back to group cultivation, worried that individuals did not distribute the multiplied seed sufficiently.

Seed crops are inspected by agency staff, often with STU participation, and advice is offered to growers on seed production. Once harvested, seed samples are sent to STU to be tested for varietal purity and germination. If passed, part of the seed crop is bought back by the NGOs but part remains with the growers. They are encouraged to dispose of the seed through informal channels—to give, exchange, or sell to family, friends and neighbours. The seed collected by the NGOs is then treated and stored centrally for distribution as registered or certified seed for the next season. SCF and FFHC are largely involved in multiplying rice seed, whilst AATG also handles groundnut and maize seed.

The GSM differs: it operates a 20 hectare farm where seed crops are grown from STU supplies and from DAR stocks of unusual seeds. Seed is then harvested, processed, and stored at the Mission. Seeds are tested both by the STU and by GSM, and are made available for purchase by applicants. However, clients are few and come mainly from within a few kilometres of the Mission at Massembeh. GSM multiplies groundnuts, maize, cowpeas, findo ('hungry rice'), and keeps small stocks of sorghum and cassava.

Servicing the NGO seed programmes, STU provides foundation seeds of just three crops: rice (eight varieties), groundnuts (two varieties), and maize (two varieties). It ignores the widely-grown crops of millet and sorghum since there are no proven varieties superior to local cultivars. It also provides services to seed multipliers in testing, field inspections, processing and advice and training. In addition, the NGO seed activities have been assisted and promoted by the USAID-funded On-Farm Seed Production (OFSP) Project, managed by Winrock (an American Foundation) from Dakar and working both in The Gambia and Senegal. This offers some valuable additional funding, technical assistance, training, and networking in seed activities for the NGOs.

To complete the picture of seed provision in The Gambia, the FAO Fertilizer Project of the Department of Agricultural Services (DAS) has, since 1988, been engaged in relatively large-scale multiplication of groundnut and maize seed, in which it encourages groups of farmers with a block of land (of 5 hectares or more) to multiply up seed crops. This is linked to a longer-standing programme of fertiliser distribution, and ties in with the Project's attempts to set up a network of private dealers in agricultural inputs, including seed.

Table 5.1 summarises the comparative performance of the main organisations involved in seed production in The Gambia. Even for a small country like The Gambia, the NGO seed programmes are small-scale and low volume: in the 1991/92 crop season they were multiplying about 50 hectares of rice, 10-20 hectares of maize, and less than 10 hectares of groundnuts. In these last two crops they were outstripped by the STU and FAO fertilizer programmes which together had 119 and 111 hectares of maize and groundnuts under multiplication.

Nevertheless, because three of the NGOs are decentralised, with seed being multiplied on small plots scattered over many villages, the coverage of these efforts and their accessibility to other farmers is greater than might be imagined. Seed growers tend to be the better-resourced local farmers, but this bias does not debar other farmers from access to better seed, although they are likely to be towards the back of the queue (none of the NGOs have investigated in-depth how the informal distribution of seed by growers works in practice). Quality of seed varies, partly because of the difficulties of standardising production in a decentralised system, and partly because of the fluctuations of rainfed farming: low or ill-distributed rainfall affects not only the quantity harvested but also the quality of seed.

Differences between crops, largely between rice on the one hand and maize and groundnuts on the other, are striking. Rice seed production is almost entirely the province of the NGOs. It is grown as a food crop by women in a variety of ecosystems, to which a profusion of local cultivars have been adapted, most of them of long (120-day or more) duration. Rice is the crop most threatened by shorter growing seasons and less rain. The NGO seed programmes have promoted new varieties of shorter duration, offered a range

Table 5.1:   Comparative performance of seed projects in The Gambia, 1991

	Seed Technology Unit	*ActionAid*	*Save the Children*	*Good Seed Mission*	*Freedom From Hunger Campaign*	*FAO Fertilizer*
CROPS, VARIETIES	Rice, 2 upland 6 lowland Groundnuts, 2 Maize, 2	Rice, 2+ Groundnuts, 2 Maize, 1-2	Rice, 6+ (Cowpeas) Millet, 1	Groundnuts, 3 Maize, 3-4 Cowpeas, 6 Sorghum, 2 Findo, 6 Cassava, 6	Rice, 4+ (local & new)	Groundnuts, 2 Maize, 2
QUALITY	Formal provision of F & R, sometimes C seeds	Variable	Good	Very good, keen interest	Not known	Not known
TIMELINESS	Delays in shipping groundnut seed	OK	OK	OK	Not known	Not known
ACCESSIBILITY	Single point of distribution, difficult communications	Decentralised	Decentralised	Little distributed; only very local, few farmers know	(Decentralised)	Changed from one area to six
PERSONS SERVED	Via other institutions	68 individual + 2 group growers	>120 growers in 20 villages	5 villages + up to 50 individuals	Not known	Not known
TARGET GROUP		Village groups	Women's groups	Mainly men	Women's groups	Mainly men
QUANTITY Area mult'd, ha Crop mult'd, ha	57 22gn, 35ma	18.5+ 7.5ri, 4gn, 7+ma	30 30ri	5.2 1.6gn, 3.6ma	10? 10?ri	189 97gn, 92ma
COSTS	US$2,592 per tonne (1984/85)	US$30,000 excl. salaries & o/heads (1989/90)	US$45,000 pa total agriculture programme	US$32,000 total Mission (1989)	Not known	Not known

Notes: FFHC and FAO Fertilizer are not properly part of this study, hence little information; gn = groundnuts; ma = maize; ri = rice; F = foundation; R = registered; C = certified of introduced cultivars for testing, and in some cases they have helped to preserve and diffuse traditional favourites. One of the NGOs, SCF, has found that the rice variety it introduced, *Peking*, has almost completely replaced the previous local varieties in parts of North Bank Division, which is a cause of concern because of the susceptibility of *Peking* to blast.

Maize and groundnuts, on the other hand, are cash crops, grown by men in more uniform ecosystems. Both crops are introductions into The Gambia, being brought from the Americas, and they have only been grown on any scale since the 19th century (groundnuts) and since the 1970s (maize). The favoured varieties are recent arrivals and, for both crops, two or three cultivars dominate. Since they have shorter durations they are less threatened by declining rainfall, so that the main motivation to produce seed is to increase yields—especially since these are the crops most likely to receive chemical fertiliser. Although AATG and the GSM produce these crop seeds, the FAO fertilizer project dominates.

Costs of formal seed production are inadequately recorded, but the evidence available suggests that NGO efforts may be costly, well above the prices paid in local markets for seed. Very rough estimates suggest an average cost of US$1.25/kg of seed produced in 1989/90, allowing for project overheads but excluding the cost of growers' labour. This is approximately four times the local market price for seed of modern varieties in The Gambia at the time. This stems from the low volumes handled compared to high overhead costs.

The NGOs have stepped into an institutional vacuum in The Gambia: government and the cooperatives are unwilling or unable to produce seeds, and traders have been reluctant to enter the seeds market. Compared to these alternatives, NGOs enjoy several potential advantages: relatively reliable funding; flexible procedures allowing response to local needs; lower transactions costs than traders, since information can be generated from their other village-level activities; and the ability to accept the losses always likely given the hazards and uncertainties of dryland crop farming. Some of the claimed advantages of NGOs may be more apparent than real: in particular, it is not clear if NGOs do operate at lower costs than other organisations.

Despite their institutional potential, and the current lack of alternatives, all of the NGOs see themselves as temporary actors in seed production. They plan for the farmers they work with to take over multiplication as a profitable activity. How and when villagers will be able to mediate with central units like the STU is not spelt out: given the poverty, rudimentary formal education, and often poor health of villagers, the prospects are not promising.

Seed is important to Gambian cropping systems and support to local supply initiatives seems warranted on three counts. First, when the rains fail, village seed supplies are often inadequate: organised multiplication and storage provides some cushioning against bad harvests. Second, seed storage is a frequently mentioned problem: efforts to improve storage should be valuable. That said, although AATG and SCF are improving village stores, storage takes second place to multiplication. Third, seed can contribute to improved cropping. At the moment the key issue seems to be defensive adaptation to declining rainfall, rather than yield enhancements. Improvements in cropping look more likely to come not from higher-yielding varieties but from attention to soil moisture, soil fertility, weed control, the grazing of stock, the use of draught animals and associated tools, and dry season vegetable gardening.

Local Seed Supply for the Hills of Nepal

Nepal is a small country (141,000 sq.km.) with a population of 18 million, of whom 54 per cent live in the hills and 95 per cent are dependent on agriculture—which provides 60 per cent of GDP. Economic growth has averaged 4.6 per cent per annum, but per capita income remains low at US$170. Nepal receives a large amount of international aid (15 per cent of GNP) and has had increasing aid absorption problems. It is recognised that local and international NGO activities need to be co-ordinated better.

Nepal has had a commitment to decentralising development since 1982. Village Development Committees have been important decision-makers, although they have tended to be dominated by local elites. The Eighth Five Year Plan (1990–95) emphasises people's participation, including privatisation, co-operative enterprise and, in agriculture, service delivery through the 'group approach'. All economic planning in Nepal is uncertain now, however, following recent political changes: the first democratic elections since the 1950s were held in 1991.

Of the total land area, some 20 per cent is cultivated: about one third is sown to maize, one quarter to rice and the rest to wheat and millet and, to a lesser extent, potatoes. Farming systems are determined by altitude, aspect and the availability of irrigation: at least 25 major cropping patterns have been recorded in the hills alone. Multiple cropping is the norm. The modal holding size is only 0.5 ha; most families get no more than six months' food from their own land each year. The key agricultural problems are heavy population pressure, low and declining yields (because marginal land is being brought into cultivation and old seed varieties are not replaced), dwindling fuel and fodder reserves and lack of irrigation and fertiliser.

There is an active farmer-to-farmer seed exchange system. Farmers use many varieties of one crop, which are obtained from different sources (extension 'mini-kits', landlords, other farmers and relatives), and they often travel considerable distances to get seed. Most transactions involve transfers of seed from richer to poorer households in the form of gifts, seed swaps, in-kind seed loans or exchanges for labour; few farmers have the resources to pay cash.

Nepal does have a formal national seed system but it is weak and has had chronic problems supplying the hills. They are inaccessible and remote and have poor seed storage facilities as well as small and scattered markets; furthermore, hill farmers want area-specific varieties. The national Agricultural

Inputs Corporation (AIC) sells around 3,000 tonnes of certified seed each year, equivalent to only 10 per cent of national demand. The government favours local level seed production and distribution as an alternative system and it is encouraging both individual private commercial traders and community-oriented local level seed production and distribution. Private traders grouped together in 1988 to form the Seed Entrepreneurs Association of Nepal (SEAN). SEAN participates actively in the national debate on seed sector development policy and in the seed training provided by the government's Central Seed Science and Technology Division (CSSTD). However, the private sector deals almost exclusively with vegetable seed and is concentrated in market centres in the plains.

The favourable policy environment for local level seed supply in the hills has led to at least 12 separate project initiatives in Nepal in the last 15 years, to replace the formal public seed sector. The experiences of three, which typify the range of approaches pursued, are outlined below.

STIP/USAID Private Producer—Sellers Project (PPS). The PPS project started under the Seed Technology and Improvement Program (STIP) in 1985 and is now part of the regular programme of the Department of Agriculture (DOA). It is an extension programme for hill farmers interested in becoming small-scale seed entrepreneurs which aims to refine existing farmer-to-farmer seed distribution systems rather than to replace them with new, large-scale formal structures. Earlier experiences in Nepal showed that large, formal systems are impossible to operate without subsidies and that it is essential to involve farmers in distribution as well as in production, if local seed supply is to be sustainable in the long run.

In each district where it operates, the project identifies five to ten potential production areas and 30–50 private producer sellers. Selected producers are often those already involved with the agricultural research services via on-farm trials. They have to be progressive farmers with larger holdings, who can bear risk and who are leaders in the local community. They are trained in seed production and distribution together with local DOA extension staff and are given metal seed storage bins developed by Nepal's Rural Save Grain project at 25 per cent subsidy.

CSSTD is the technical agency responsible for planning, co-ordination and technical support; AIC delivers foundation seed supplied by CSSTD; district-level implementation is through DOA staff, who supervise growers and help them with foundation seed, treatment and promotion (there have been some problems with co-ordinating the different line agencies involved.) The growers are responsible for all other activities, including seed sales.

The seed crops grown are modern varieties of wheat, rice and maize; the project is also diversifying into vegetable seed.

Visual quality assessment is carried out in the field by DOA staff and germination tests are also carried out locally. Seed quality is apparently good.

Seed is stored on-farm by individual farmers: the project deliberately works with individuals rather than groups, in order to keep the price of seed low by avoiding transport to central storage points. However, only half the seed produced is stored in the metal bins provided.

By 1988, the project was producing 110 tonnes of certified wheat, maize and rice seed. The wheat seed was 25 per cent of the requirement in the project area at the time, assuming farmers buy fresh seed every seven years. It was 50 per cent more than the previous AIC supply to the area (AIC does not sell its own subsidised certified seed in areas where the project is operating, to prevent unfair competition). The target for 1991 was 320 tonnes and PPS seed now accounts for 5 per cent of all certified seed used in Nepal; three of the districts are producing enough to export seed to other areas.

It was originally planned that the project would cover 18 to 20 of the 40 hill districts. After starting in four districts in the west, it has extended to three more, but the project can operate only in areas with readily available inputs. It has trained 1,300 farmers and 40 extension workers and there are now some 300 private producer-sellers. Although women play a major role on-farm, it is difficult to get them to participate in off-farm activities such as training because of cultural norms. As a result, only 25 per cent of the official participants are women. The incentives to growers are priority allocation of fertiliser, pesticides and research packages of new varieties, equipment and other support.

Growers set prices with DOA staff supervision. The minimal capital investment and low transport costs help to keep prices down, usually to not much more than local grain prices. This contrasts with AIC seed prices, which are 50–90 per cent above grain prices, depending on the crop. However, the cost to the project of CSSTD's co-ordination role, DOA's supervision, the seed bin subsidy, the transport of foundation seed by AIC and staff and farmer training are not included in the pricing calculation. The budget for these activities in 1991 was US$5,300, equivalent to US$0.02 per kg of seed produced (Dickie, pers. comm.).

Koshi Hills Agriculture Project (KHAP).   KHAP is part of an area development programme operating in the eastern hills of Nepal (6,500 sq.km.; population 590,000) since the late 1970s with UK ODA support. Since 1987, it has operated a seeds programme, originally established under the Pakhribas Agricultural Centre (PAC), the local agricultural research station. The dual aim of the seeds programme is to increase the availability of improved seeds in the Koshi Hills and to improve farm incomes by establishing local seed producer groups (SPGs). The agriculture project also includes extension, training and building improvements and revolving funds for local agricultural service centres (ASCs) and co-operatives.

The group approach has become an important part of the strategy. Farmers who meet the selection criteria are organised into SPGs and then trained in seed production, supervised and provided with help with input procurement problems and with marketing. This degree of support for the SPGs is felt to be necessary to compensate for national problems with input distribution and also because many groups have been formed only recently.

The programme is supposed to be fully integrated into the local DOA programme but the planned government input of staff has not been provided. KHAP has had to take responsibility in most areas; nonetheless, liaison with local extension staff is good. PAC has also had a large input, both in concept development and in continuing field inspection and laboratory sampling (PAC is a regional seed testing centre for CSSTD). KHAP staff display a high degree of technical competence.

The first seed crops produced were rice, wheat, maize, millet and potato (the focus crops of the government's Basic Needs Policy); subsequently KHAP has expanded into soyabean, lentils and vegetables. Low-mid altitude crops and varieties dominate, because these are the ones for which new varieties are available and willing growers exist.

There are now 32 SPGs with 650 members; 45 per cent are men and 70 per cent are high caste. Between 1987 and 1991, 235 tonnes of seed were produced, providing 40–60 per cent of all the seed supplied in the Koshi Hills. Recent estimates suggest that this has provided up to 20 per cent of the population in the programme area with some two months' extra food supply and a 10 per cent increase in annual cash income, depending on the balance in seed production between cash and food crops. But this benefit will have been skewed in favour of farmers who can pay cash for seed and who can achieve the potential incremental yield of the low-mid altitude varieties produced by KHAP.

Each SPG is associated with a co-operative responsible for selling seed in the local area, purchasing it using a revolving fund provided by the project. KHAP staff also move surplus seed to other areas and to local private traders. Seed selling prices for growers are supposed to be 10 per cent above local market prices at harvest plus a 20 per cent premium for seed passing all laboratory tests. Co-operatives sell seed at 10 per cent above this buying price plus handling and storage expenses; this is supposed not to be more than the local market price at planting time.

Growers buy all the inputs they need at normal prices. They sell seed at a price fixed locally by the co-operatives, DOA and project staff and growers. As on similar projects elsewhere in Nepal, returns are not significantly higher than for grain production, due to higher labour inputs, low seed yields and the small seed/grain price differential. The main advantage for growers is better access to inputs and agronomic advice.

All the overhead costs of staff salaries and staff transport costs are met by KHAP. These totalled US$766,000 in the three years from 1987 to 1990 or US$3.25 per kg of seed produced. Seed certification services are provided free by PAC. The DOA is supposed to provide 40 per cent of the programme budget but so far has not been able to do so.

ActionAid–Nepal (AA–N). Since 1982 AA–N has been involved in a number of community programmes, including agriculture, in Sindhupalchowk District, central Nepal. One of the objectives of the agriculture programme is to increase yields sustainably and AA–N's Seed Production Programme, set up in 1985, contributes to this by ensuring self-sufficiency in seed of the most successful varieties while allowing local farmers to earn income from seed production. The agriculture programme also includes research and extension on new varieties, pest and disease control, irrigation, agricultural education and agronomic practices.

Local committees choose the area and seed crop and set up seed producer groups. Growers tend to be higher income households as poor families do not have enough land to allocate to seed production. AA–N provides agricultural assistants who supervise production and training, and provide seed (on an exchange basis) and loans for fertiliser. Fertiliser loans are being phased out and the project is now promoting biological fertiliser, pest control and storage treatments. Metal seed bins are provided at a 25 per cent subsidy. An AA–N agronomist and agricultural programme planner also assist the programme.

Training is provided for the whole household, to encourage women to participate. As well as seed production and storage it covers how to establish linkages with outside agencies for foundation seed, with the aim of making the groups self-sustainable.

Quality problems with initial seed supplies have made AA–N change to supplying only seed from government farms, which is then multiplied up in the project area. As there are problems with farmers mixing and damaging seed, production targets are deliberately set higher than needed. The project does its own germination tests.

Seed crops grown are, in order of importance: rice, wheat, maize, millet, barley, soyabean and potato. AA–N has carried out a significant amount of research into the best local and modern varieties for area.

The project cultivates about seven hectares of cereal seed and 0.3 ha of potato and vegetable seed per year, all of which is distributed within the project area: committees are now self-sufficient in wheat, maize, barley and finger millet seed. In 1990/91, of the 6,400 households in the project area, some 730 used project produced seed and 1,210 used seed bought in by AA–N and distributed at subsidised prices. The intention was to increase the number of households benefitting from project-produced seed to 1,000 in 1991/92. AA–N recognises that the seed programme is of limited relevance to the poorest families, as they have little land and their main need is to put labour to productive use off-farm.

Seven farmer households have been trained so far. They are given source seed free but return an equivalent amount to AA–N at harvest. They produce on contract for AA–N and are encouraged to go on to grow for AIC and private companies. At first, AA–N bought all seed and was responsible for storage and distribution. Now, farmers are encouraged to sell seed themselves, with support from AA–N in marketing and an advance to allow farmers to store over the off-season and sell at higher prices at sowing time (AA–N still sells the seed it receives back from growers as payment for source seed). Seed prices are based on prevailing market prices plus a premium, so they are lower than AIC prices but higher than local grain prices; there have been problems with growers wanting higher prices.

AA–N expenditure on the agricultural research and extension programme, which includes the seed programme, doubled during 1990/91 to US$23,000; seed production is one of eight sub-components. Gross seed project costs were estimated at US$1.20 per participant or US$0.54 per beneficiary.

The aim of encouraging local seed supply for the hills of Nepal is to reduce production costs and organisational complexity and to increase economic efficiency, to make appropriate varieties available on time, to reduce transport costs and retail seed prices and to maintain quality standards. There is no obvious difference in the experience of the three different types of project: these projects, and others like them in Nepal, are now making a significant contribution to the availability of improved seed in the hill areas (particularly compared to the performance of AIC) and have proved beyond doubt that local farmers can produce seed of good quality. However, the availability of appropriate varieties to multiply for the hill areas depends on the output of the national agricultural research system, which is still not well-oriented to the needs of hill farming systems.

One important wider issue is the long-term benefit of this type of local seed supply for growers and users. The limited cash benefit of growing certified seed is limited and the technical and organisational requirements of seed production mean that the poorest farmers cannot be involved. And for users, the need to pay cash precludes the poorest households from using seed produced by projects such as these and there are wider questions about the relevance of modern varieties for households with little land and poor access to complementary inputs. In addition, all the projects reviewed here have substantial overhead costs which are currently absorbed by donor funding. This relates to the main question facing local seed supply in Nepal at present: how sustainable are the initiatives without continued project intervention in terms of budgetary and institutional support?

6
National and International Seeds Advocacy

NGOs and International Seeds Advocacy

NGO efforts to strengthen the local supply of both appropriate modern varieties and landraces continue in many individual projects and programmes in Asia, Africa and Latin America. However, these initiatives face mounting competition from multi-national and other interests, seeking to increase the number of small farmers in developing countries using conventional high-yielding varieties and the accompanying packages of fertiliser and crop chemicals developed under the protection of plant breeders' rights and patents. Over the last decade, a number of NGOs have seen that maintaining sustainable agricultural systems on individual farms in the developing world depends increasingly on raising awareness internationally of the global dangers of declining genetic diversity and of the inequity of ignoring farmers' rights over indigenous genetic resources. In this chapter, we profile the experiences of three international campaigns which have advocated farmers' rights to appropriate seed to policy makers and the wider public.

Genetic Resources Action International. GRAIN is a small independent NGO operating out of Barcelona, Spain, with three full-time staff and an annual budget of US$118,000. It is funded by other NGOs, mostly European, and by European government agencies. It acts as the European contact point for Seeds Action Network International, which comprises in addition RAFI for North America, ELC in Kenya for Africa and SAM in Malaysia for Asia.

GRAIN developed out of the International Coalition for Development Action (ICDA) Seeds Campaign, which operated from 1975 until GRAIN was set up in 1990. The campaign aims to stop national legislation granting monopoly control over plant genetic resources and promote the adoption of international agreements, under the UN, to regulate the exchange and promote the conservation of genetic resources. GRAIN works to promote popular action against genetic erosion, which it sees as one of the most pervasive threats to world food security, undermining sustainability and eroding the options for development. It was launched in the belief that European NGOs play a vital role in stimulating popular action and policy change towards an improved 'genetic resources order', both within Europe and internationally, but that they require a means of more effective networking and co-operation in order to achieve this.

GRAIN's work is organised around raising public awareness of genetic erosion; increasing knowledge and understanding of its causes and its implications for the poor; stimulating activities and policies for the conservation of genetic diversity at local, national and international level; and supporting the activities of individuals and groups concerned about these issues and facilitating co-operation between them. Its core work programme consists of a regular newsletter, *Seedling*; books and conferences on genetic resources issues; and work with international organisations such as FAO, contributing to international agreements concerning genetic resources. It also has a series of special projects which, for 1991–93, consist of: *Fight For Rights* (high level political lobbying); *Conservation in the South* (direct collaboration with community organisations in the South); and a campaign to increase awareness of the role of NGOs in maintaining the genetic base of European agriculture. ·

The main successes of the Seeds Action Network International, in which GRAIN has an important role, are: achieving greater global awareness of, and consensus on, the importance of plant genetic diversity for world food security—for example, acceptance of the principle of Farmers' Rights by multi-lateral agencies; issues of access to plant genetic resources were incorporated into the UNCED Biodiversity Convention; and achieving greater global awareness of the problems surrounding patenting life forms (for example, the recent rejection by the European Parliament of the EC Commission's proposal to proceed with legislation permitting the patenting of life forms within the EC).

The major factors hampering GRAIN's work at present are the increasing privatisation of agricultural research and plant genetic research products, and the lack of good data on the performance of landraces and farmers' varieties compared to HYVs and other MVs.

Rural Advancement Foundation International. RAFI was also originally associated with ICDA. It revolves around a team of seven staff, divided between offices in the United States, Canada and Australia and has an annual core budget of US$600,000 provided by Canadian CIDA. The staff started working together on seeds issues in the mid–1970s and achieved international prominence with publications such as *Seeds of the Earth* and *Law of the Seed*. By 1984, RAFI was incorporated in The Netherlands as a non-profit NGO. It focuses on publicising the socioeconomic impact of new agricultural technologies on rural societies and especially the global political issues associated with the conservation and utilisation of plant genetic resources, including the maintenance of genetic diversity and the impact of biotechnology. It is independent of any single country and organisation and targets international organisations such as FAO, GATT and WIPO. It also works with community-level grassroots organisations, as a consortium with its partners CLADES in Latin America, SEARICE in Asia and ASS in Africa.

RAFI's broad mandate is to carry out research, education and advocacy work to make the issues surrounding international plant genetic resources widely understood and to act as an early warning system for important developments that might adversely affect farmers' access to plant genetic resources. It does this through four main areas of activity:

- **publications**: of popular books on seeds issues and of regular printed and E-mail newsletters (*RAFI Communique, RAFI Express*);
- **lobbying**: of international, particularly UN, organisations dealing with plant genetic resources issues;
- **supporting developing country partners**: by providing information, including international, regional and national workshops, and financial support for partners implementing plant genetic resources conservation;
- **community seed bank kits**: in 1986 RAFI pioneered a kit which provides all the information necessary for NGOs to set up community seed banks in developing countries and to undertake plant breeding for conservation and development of landraces.

All RAFI's activities involve inter-regional South–South and South–North co-operation and are targeted at programmes of practical action.

Despite problems in obtaining funding, particularly for infrastructure for its regional partners, RAFI has had considerable success in putting pressure on international organisations to take account of the plant genetic resources needs of farmers in developing countries. For example, together with other international NGOs, it was active in promoting the FAO International Undertaking on Plant Genetic Resources, the first international agreement to promote the conservation and free availability of genetic resources for plant breeding. RAFI was also instrumental in promoting the concept of Farmers' Rights.

African Seeds of Survival. ASS is one of the collaborative initiatives to which RAFI is contributing. It is a large integrated programme of community, national and regional genetic diversity development operated out of Ethiopia by a consortium of Canadian NGOs, led by the Unitarian Service Committee (USC) and funded largely by Partnership Africa Canada. ASS started in 1988 with US$1.1 million funding for 1988–91, now extended to the end of 1992 and with an anticipated extension after that.

The key distinguishing feature of ASS is the combination of practical support for community-level *in-situ* genetic resources conservation in a developing country with awareness raising about plant genetic resources issues both continent-wide across Africa and in secondary schools and agricultural colleges in Canada. Eleven constituent sub-projects are organised around four main activities:

- **farmer-based seed improvement**: a farmer/breeder co-operative initiative to restore landraces lost in Ethiopia during the 1980s droughts and to select from them to increase yields; training of NGO rural extension

agents from Africa in seed collection, storage, growing-out and documentation, at the Ethiopian gene bank (PGRC/E); on-site advice for NGOs from PGRC/E technicians on seed-saving and plant-breeding;
* **network development:** a pan-African organisation and three African regional programmes to address the problems and potential of biological diversity; and funding and implementation support for an African Commission on Biological Diversity linking governments, scientists and NGOs, to address the socio-economic impact of the new biotechnologies on biological diversity in Africa;
* **information projects** to support farmer-scientist co-operation in breeding: provides a Newsletter; a community plant breeding kit for NGOs; and a loose-leaf binder to be up-dated annually covering information and assistance sources for plant genetic resources issues in Africa, by crop;
* **development education** in Canada: an Ethiopia exchange programme for Canadian farmers; a national conference in Canada on biological diversity; and a development education kit for Canadian secondary schools and agricultural colleges.

Each of these was initially proposed by a different organisation, mainly African. USC is responsible for the overall management but numerous organisations contribute, including PGRC/E, ENDA–Zimbabwe, IFOAM and AAASA.

The farmer/breeder initiative is organised by an ASS agronomist in the field and involves the Ministry of Agriculture, the PGRC/E and Ethiopia Seeds Corporation (ESC) as well as farmers from 18 volunteer farmers' associations in two administrative regions of Ethiopia. Source seed of 14 landraces of five locally important crops has been obtained from farmer collections and from ESC and given to local farmers to multiply up. At harvest, farmers return an equivalent amount to ASS and can also sell any excess to ASS at local market prices. ASS stores the seed and distributes it again the following season for further multiplication. ASS has also provided a truck and arranged the farmer contracts. The intention is that, in time, sufficient seed will be available locally for maintenance to be assured simply through self-sustaining local propagation and ASS can withdraw. In the short-term, ASS hopes to attract other NGOs in Ethiopia to the initiative.

So far, the initiative has involved some 27,000 farmers with barley, chick pea, faba bean, sorghum and teff seed and the number of landraces has increased from 14 to 18. Elite wheat lines are also being maintained for a university breeding programme. In 1989/90, the latest recorded season, 67 tonnes of seed was produced on 40 hectares. At present, most of this is still used for further multiplication so there has not been any widespread distribution for use as crop seed yet. The number of sites has increased from 18 to 21 (36 sites are planned), all in drought-prone areas. Only farmers in farmers' associations (about two thirds of all farmers in Ethiopia) are eligible to participate.

The farmer/breeder initiative takes just over 50 per cent of the total ASS budget, estimated at US$3.24 per kg of seed produced, making possible an average yield increase of 5 per cent per year per household. Half goes on seed cleaning, etc. and farmer contracts, nearly one quarter on the agronomist services and the rest equally on vehicles and administrative support. In comparison, networking takes just over 20 per cent of the budget, information projects less than 5 per cent and development education work just under 10 per cent; the remainder is used for programme development and administration.

The de-collectivisation of agriculture in Ethiopia in 1991/92, and the destabilisation caused by the intensifying war and subsequent change of government, caused some problems for the initiative and it is now running about one year behind schedule. The extension agent training has got well under way, with two courses, each for around 12 African participants, each year. Each course costs in the region of US$16,000. The development education work has been expanded to include African as well as Canadian schools and colleges. However, it has been delayed by contractual problems with the production of teaching materials. The exchange programme has been suspended.

Despite these problems, all the ASS programmes continue and have attracted considerable attention as an alternative model for plant genetic resources conservation work. In 1989, Melaku Worede, the PGRC/E Director, was awarded the Right Livelihood Award (the 'alternative' Nobel prize) for his plant genetic resources conservation work and immediately donated his prize money to ASS. At the time of writing, a similar programme has been devised for Asia and it is hoped that an *Asian Seeds of Survival Programme* will soon be operational.

Towards Sustainable Local Seed Supply in The Philippines

The Philippines is a densely populated archipelago in South East Asia, with a population of 60 million covering 300,000 sq.km. Agriculture provides the livelihood for 70 per cent of the population and contributes 22 per cent of GDP. Although economic prospects were good 20 years ago, growth is now less than 2 per cent per year and rural poverty is increasing, mainly due to continuing uneven access to land (85 per cent of farmers are landless tenants). The agriculture sector is facing a crisis caused by increasing input costs, declining yields, increasing disease problems and ecological damage. The Philippines became a net food importer again in 1989.

Nearly 30 per cent of the total land area is cultivated and rice dominates; maize, coconuts, vegetables, other cereals and legumes are also grown. The average holding size is only one hectare and cropping is intensive, with up to three crops per year. The Green Revolution in the 1960s had a big impact in the main lowland irrigated rice area but maize continues as a subsistence, unimproved crop grown on marginal land. Improved legume varieties are available but demand for them is low. Only about 10 per cent of farmers use certified seed because the potential incremental yield is declining and requires large inputs of fertiliser, and most farmers have to borrow cash from local money lenders at high interest rates to pay for this package.

About 152,000 tonnes of certified rice seed and 74,000 tonnes of certified maize seed is needed annually at recommended replacement rates but less than 10 per cent of this is produced at present. The formal seed sector in The Philippines was established before the era of large, donor-financed, national seed programmes. For rice, it is based around small-scale farmers using family labour and traditional, labour-intensive production methods. There are some 72 Seed Growers Associations with 1,500 members that operate without subsidies, supported by regional seed testing laboratories. Seed growers are relatively larger and more prosperous farmers (the average holding size of members is 7 hectares). Truth-in-labelling seed legislation allows seed to be sold direct from these farms. There is also a substantial multi-national and local commercial private seed sector, dealing mainly in maize seed but also in vegetables.

The Bureau of Plant Industry is the lead government agency in seed production and distribution and is responsible for research stations and seed services. The Philippines is well-equipped with formal plant breeding facilities at the university, IRRI and Philrice but there is little farmer participation (on-farm trials are carried out only for approved varieties) and the emphasis is still on high-input varieties for maximum yield.

Government policy has had a significant impact on the uptake of HYVs in the past by including them in subsidised agricultural programmes. It is still official policy to encourage production and distribution of modern varieties rather than conservation of traditional varieties. In addition, there is now pressure from some international aid donors, notably USAID, to introduce patents, to liberalise seed imports and to encourage greater private sector involvement in the domestic seed trade, in the belief that this will stimulate the development and availability of modern varieties.

At the same time, present government policies recognise the significant role NGOs play in the development process and there are over 100 NGOs and farmers' organisations involved in local seed supply, all focusing on the conservation of farmers' traditional varieties. There is a national network of NGOs (SIBAT) which supports sustainable grassroots development and is particularly strong on traditional rice varieties, and an umbrella group of farmers' organisations and national scientists that is supporting the development of an indigenous rice industry. In the 1980s, several farmers' organisations launched a special coalition (MASIPAG) specifically to develop low-input rice varieties.

Mindoro Institute for Development Inc (MIND) is a member of SIBAT. It started development work in the province of Occidental Mindoro in 1984, as an off-shoot of the Farmers Assistance Board, a Philippines NGO which organises extension services for peasant farmers. MIND is the main NGO in Occidental Mindoro, operating a comprehensive integrated rural development programme including participatory agricultural, health, education and legal services. The preservation of genetic resources in rice farming, especially the identification and promotion of alternative low-input varieties, is central to MIND's strategy for sustainable agriculture, which forms a major part of its development work.

Occidental Mindoro is one of the two provinces on Mindoro, the seventh largest island in The Philippines; it has a population of 290,000 and covers 588,000 ha, of which 17 per cent is cultivated. Rice monocropping is the main occupation of 80 per cent of farm families, most of whom are subsistence farmers: few other crops are possible because of flooding in the main cropping season and shortage of irrigation water for the second cropping season. Land reform is planned for over one quarter of the cultivable land, but little has been implemented so far.

Rice yields average 3.3 tonnes per ha, which is considered low, and most official development effort is concentrated on increasing yields to the 4.5 tonnes per ha target set by the government. Many traditional varieties are being lost because of this intensive cultivation: 85 per cent of Mindoro's rice area is covered by just five IRRI varieties, although only about 6 per cent of the rice area is planted with certified seed each year.

In 1988 MIND established a community seed bank (CSB), linked to two demonstration farms, to test local traditional rice varieties, to make them accessible to farmers, to teach rice breeding to farmers and to promote sustainable, environmentally sound agricultural methods. In 1989, the Institute received a grant and training from SIBAT for the CSB, and help with information exchange with other interested NGOs. MASIPAG has also helped with a number of varieties and staff training.

The first CSB started close to Occidental Mindoro's main commercial district. There are plans to start three more within MIND's project area. The project was reorganised in 1991 and the CSB is now controlled by a community-led Livelihood Enhancement Committee (LEC), to make it more participatory.

The CSB and associated farms are managed and worked by LEC members on a rota basis: they choose what varieties to grow and they provide all the labour (1–2 days per week each). A strong community organisation is seen as essential for the success of the CSB, not least because of the large amount of labour required for maintaining all the different varieties properly.

The CSB works mainly with rice, although the project hopes to expand into vegetable seed and legumes. It started with 40 varieties of rice, mostly farmers' own varieties but also some modern ones, including some of IRRI's high-yielding varieties. It now has 180 varieties, all farmers': 130 were donated by MASIPAG; 40 came from SIBAT's seed exchange network; and 10 were obtained from local collections.

Germplasm collection, variety evaluation and variety maintenance is done by four agriculturalists, three local and one expatriate volunteer. In addition, SIBAT provides technical assistance and a group of scientists from the university act as MIND's technical consultants. The donated varieties are put into field adaptability trials to obtain varietal characteristics and yield data. Variety evaluation is done during the annual farmers' field day, by visual inspection of standing crops and taste tests. Once varieties are proven to be suitable for the area, they are multiplied up. Since the project started, 10 farmers' varieties have produced good yields at low input levels (up to 3.3 tonnes per ha), comparing favourably to average national fertilised yields. They have good culinary qualities as well.

Source seed for multiplication comes from the CSB's own maintained stocks. Seedlings are started in the nursery on raised beds, to prevent snail damage, and then transplanted using a new technique, which reduces weeding by up to 40 per cent and facilitates snail control by ducks. After harvest, the seeds are sun-dried, winnowed and then stored in labelled sacks (or plastic bottles for small quantities) in a rat-proof store. Storage conditions mean that the seed is viable for only one year so it is always grown out the following season. There have also been problems with seed purity due to the large number of varieties being multiplied.

Forty-five families belong to the LEC that is involved with the CSB and most work on the project farm. Once sufficient seed of a proven variety has been multiplied up at the farm, farmer members can take seed of the varieties of their choice at no charge, in quantities proportional to their labour input. They are also given training in sustainable, environmentally friendly agriculture. Non-members can take 2–3 kg seed at planting time and return the same quantity after harvest (they can take 25 kg plus but in this case they have to return an additional 10 per cent). Members are also encouraged to share seed informally with their neighbours and other farmers.

Through this distribution system, all LEC members are now using seed from the CSB and some 400 of the total 1,280 farmers in the MIND project area have also received seed (now estimated to cover up to 600 ha of local rice land). Farmers in a number of other areas in Occidental Mindoro have also obtained CSB seed but there has been no formal investigation of the extent of this spread.

Farmers planting CSB seed can make considerable savings when it is combined with the production practices recommended by MIND (for example, substituting rice straw for NPK fertiliser and animal manure for urea). In addition, less labour is required for harvesting, (although more labour is needed for hand weeding and snail picking, to replace pesticide applications). If the price of local rice varieties reflected farmers' preferences for them, the returns to using CSB seed could be up to 30 per cent higher than returns to the prevailing modern varieties. Nonetheless, returns to the standard high-yielding variety package remain higher in the absence of this price premium. Although farm families prefer traditional varieties, there is no premium so far because they are not yet popular with private traders and The Philippines National Food Authority, who purchase most of the rice produced in Mindoro at around US$0.14 per kg.

Cost data for producing CSB seed have not been recorded (all operating costs are covered by MIND) but it is known that costs exceed prevailing prices for certified rice seed produced by the formal seed sector. Although seed prices are in theory government-controlled (rice seed officially cost US$0.39 per kg in 1991), seed prices charged by traders can be much lower in practice because the formal sector seed production system is very low technology and growers can discount retail prices if they wish.

7

Performance Assessment

In this chapter, we compare the effectiveness of the different seed activities profiled in the case studies according to a range of performance indicators, taking into account the assessments of the agencies themselves and, where available, of the farmers they seek to serve. Indicators of performance are grouped into five categories, based on the list of key seed needs given in Chapter 2:

- **technical**: the appropriateness of the seed produced for local needs in terms of variety and quality. Questions of quantity and scale of production are important here;
- **managerial**: do the agencies organise and operate effectively, and what are their strengths and weaknesses? Questions of timeliness of delivery and accessibility are important here;
- **economic**: do the agencies' seed activities produce a benefit greater than the cost of the resources employed?; what are the benefits and the costs?; are farmers prepared to pay for the benefits?;
- **social**: do the agencies' seed operations improve or worsen welfare?; do they help the poor and marginalised?; what is their impact on distribution?;
- **institutional**: what is the contribution of the agencies' approach to seed production and distribution to the sustainability of local seed systems over the longer-term?; what factors could lead to the abandonment of the seed activities?

This last indicator should reveal a critical difference between the type of approaches used by NGOs and those used by conventional seed projects and programmes. It is one of our main concerns here. For any seed system to be sustainable without external support over the long run, two conditions have to be met. First, the system must be able to operate without institutional support from external agencies (for example, in organising operations or as advocates for the community to formal sector services). This usually requires a high degree of community participation in the seed activities, and it requires ultimate control of the seed system to rest with the community. Second, the system must be economically viable. Simplistically, this means that the benefits the seed provides to farmers who use it must be large enough to ensure that demand for seed is sufficient to cover production costs. This must be attainable in the long run without subventions from external agencies: the full cost of the seed activities (the 'economic' costs) must be covered by the prices charged for seed (although it is acceptable for payment for seed to be made in other means of exchange besides cash).

Box 7.1 summarises the performance of the agencies profiled in Chapters 3 to 6 according to the five categories of indicators. The overall performance of the case studies is then assessed using these indicators, and the relative performance of NGOs compared with that of other types of agency.

Choice of Crops and Varieties

In theory, agencies concerned with strengthening local seed systems should be particularly concerned with the appropriateness of the varieties produced–in terms of input requirements, time to maturity, taste, etc. In most situations, this will necessitate a concern with maintaining a broad genetic base in the local farming system, which incorporates landrace material and farmers' varieties as well as modern varieties developed by the formal seed sector.

All the agencies emphasise seed for food staples. However, most offer (or have plans to offer) a portfolio of crops and varieties, including some cash crops. ACORD in Mali, MIND and, indirectly, OXFAM have aimed to provide suitable material by dealing only with farmers' varieties but fully half of the agencies concentrate on modern varieties. A third specifically mention having had problems because the varieties used were not those preferred by the farmers in the project area: either farmers wanted to use only local varieties; or the project provided local varieties but not the preferred ones; or farmers wanted modern varieties. The agencies' experiences clearly demonstrate the practical problems with incorporating plant genetic resources' conservation strategies into dynamic farming systems. SCF was faced with the dilemma of seeing just one of its introduced rice varieties almost completely replace the range of local varieties previously grown, which were no longer suitable because of declining rainfall.

Often the choice of varieties has been dictated by factors external to the agencies, such as the output of agricultural research institutions. In The Gambia, for example, where early-season millet is the first or second most important crop in many villages, millet seed was not included in the agencies' programmes because the formal research system has no improved varieties for millet.

Only three of the agencies, GPSN, MIND and ASS, do breeding work (all in the form of mass selections); most use material from the national agricultural research systems or existing local varieties. However, the agencies in The Gambia and Bangladesh have brought in material from other countries and tried it out.

One third of the agencies do not test varieties for local adaptability before promoting them, or else satisfy themselves that this has already been done by another agency; they simply take nationally released varieties and multiply them up in the project area.

Seed Quality Control

Nearly one third of the agencies do not use formally accredited source seed; most of these are relief-oriented agencies working through seed banks forced to do this because of lack of alternatives.

The importance attached to quality control varies between agencies. About half assiduously harvest, treat, store and test the seed they produce (although only MCC packs, labels and brands its seed). They obtain very good test results and would have little problem in getting the seed formally certified. In most of these cases, high seed quality resulted from central control of the key quality control operations. CESA is an exception which has deliberately rejected formal seed quality control (i.e. International Seed Testing Association methods and standards) in favour of 'artisan' seed quality standards defined by CESA itself. These appear to give satisfactory results.

A quarter of the agencies are, at least theoretically, linked to formal seed quality control systems and various government agencies are supposed to test and certify project seed. For a number of the projects, this has been problematic in practice. In Nepal, Mozambique and The Gambia the government seed testing authorities do not have sufficient resources to reach all the scattered production plots in order to carry out the necessary field inspections and laboratory tests in time for the seed selling season. This has reduced the ability of the projects to offer quality seed for sale in good time for planting.

The other agencies accept lower standards or do not have any formal systems for quality control. Their seed is generally acceptable. However, overall nearly half the agencies mentioned that problems with quality had occurred at some time, often due to the initial ignorance of project staff about the quality control measures needed for seed (see below, p. 87).

Quantities of Seed Produced

The relevant measure of quantity is a comparison of the quantity of seed produced with the amount of seed the community can use. This will depend on the area covered by the project and the importance in the local farming system of the crops included in the project (for niche crops, such as certain horticultural crops, relatively small quantities of seed may be required). Some comparison with the quantities available from other potential sources of supply, such as government agriculture departments or national seed companies, is also relevant.

MIND is the smallest on-going project: it produces just 5 tonnes of seed each year, enough for around 100 hectares also relatively small in scale. The largest operation is Concern's, which provided 870 tonnes of seed for some 68,000 households in 1991. The other relief-oriented projects are also comparatively large-scale. Most of the other projects also provide substantial amounts of seed: over 50 tonnes of seed per year, sufficient for 1,000–6,000 households and more if prevailing small farmer seed replacement rates are taken into account.

Box 7.1   Case Studies: Summary of Performance

Agency/Country	Date started	Main Crops	Varieties	Quantities tonnes/p.a.	Quality	Price
ACORD, Mali	1976	Sorghum, millet	Local	460	No formal control	Capital costs absorbed, local prices + overheads
ACORD, Sudan	1989	Legumes, sorghum, millet	MV*, local	0.5	Dressing & careful storage	Joint funding, local prices
OXFAM, Sudan	1985	Millet	Local	40	No formal control	Self-financing, local prices + overheads
CONCERN, Sudan	1991	Sorghum, millet, sesame	MV*, local	870	Purchased outside area	Free
CIC, Mozambique	1988	Maize, food staples	MV*, local	185	Certified	Overheads absorbed, national prices ($0.52)
CESA, Ecuador	1989	Potatoes	MV*, local	30	'Artisan quality'	Sold at cost ($0.11) after free distribution
MCC, Bangladesh	1981	Soyabeans, vegetables	MV*, local	50, 0.39	Strict quality control	Overheads absorbed, subsidised ($0.30)
ActionAid, The Gambia	1985	Rice, maize, groundnuts	MV*	32	Some certification	Overheads absorbed, local prices ($0.30) + premium
SCF, The Gambia	1985	Rice	MV*	39	Some certification	Overheads absorbed, local prices ($0.38)
Good Seed Mission, The Gambia	1985	Food staples	MV*, local	14	Strict quality control	Overheads absorbed, local prices ($0.50)
FFHC, The Gambia	1985	Rice	MV*, local	13 (est.)	Not known	Overheads absorbed, local prices ($0.38)
FAO, The Gambia	1988	Groundnuts, maize	MV*	240	Certified	Overheads absorbed, local prices ($0.25)
PPS, Nepal	1985	Wheat, rice, maize	MV*	110	Certified	Overheads absorbed, local prices ($0.24)
KHAP, Nepal	1987	Food staples, vegetables	MV*	118	Certified	Overheads absorbed, local prices + premium ($0.24)
ActionAid, Nepal	1985	Food staples	MV*, local	7	Germination tests by project staff only	Overheads absorbed, local prices + premium ($0.26)
ASS, Ethiopia	1988	Food staples	Local	67	Not known	Overheads absorbed, local prices
MIND, The Philippines	1988	Rice	Local	5 (est.)	No formal control	Overheads absorbed, seed loans

Notes: Date started = present seed project only; prices = US$ per kg of major crop produced by project.

Source: Case studies

Box 7.1 Case Studies: Summary of Performance *(continued)*

Agency/Country	Production system	Distribution system	Local involvement	Sustainability
ACORD/Mali	Farmers return loaned seed + 50%	84 village seed banks	Village management committee	Threatened by harvest failure and civil insecurity
ACORD/Sudan	ACORD trials and other seed projects	4 village seed banks	Village management committee	Lapsed after harvest failure and refugees left area
OXFAM, Sudan	Farmers return loaned seed + 20-40%	16 village seed banks	Central management committee of villagers and project staff VRCs distribute seed	Threatened by harvest failure
CONCERN, Sudan	Purchased from local merchants and NSA	Via Village Relief Committees		Relief operation
CIC, Mozambique	Project contracts local farmers	Local cash sales by contract growers	Project employees own shares; war prevents wider involvement	Threatened by war and problematic wider links
CESA, Ecuador	Community seed plots	Communities keep seed for own use and sale	Collaborative venture with local farmer federations	Recent venture but potential as an alternative model
MCC, Bangladesh	Contract growers, purchased from local wholesalers	Cash sales by project within and outside area (9,000 families)	Minimal	Project hopes farmers and local dealers will take over but high staff input to date
ActionAid, The Gambia	Project contracts village groups	25% of harvest kept by growers for our use and distribution; 75% kept by project as source seed	Village groups decide crops and individual growers	Project anticipates groups will take over within 4 years
SCF, The Gambia	Groups and individuals selected by village development committees	Seed loan returned to project; growers keep remainder for own use and distribution	Village development committees select growers	Project hopes farmers and entrepreneurs take over within 3 years but high staff input to date
Good Seed Mission, Gambia	Hired labour on mission farm	Sold by mission to circa 50 selected farmers and institutions	Minimal	High level of mission involvement to continue
FFHC, The Gambia	Project contracts groups and individuals	Project buys back other varieties; farmers distribute other seed	Problematic	Project approach being developed: successful contract system not identified
FAO, The Gambia	Block demonstration groups contracted by local dealers	Cash sales by dealers	Project liaises contracts	Project hopes to withdraw to service support; threatened by harvest failure
PPS, Nepal	Individuals selected by project	Cash sales by growers	Individuals rely on project	Project hopes to withdraw within 4 years; problematic wider links
KHAP, Nepal	Groups selected by project	Cash sales by co-ops and growers	Groups rely on project	Project hopes to withdraw within 3 years; problematic wider links
ActionAid, Nepal	Groups selected by local committees	Cash sales by growers now being encouraged; project also distributes (730 families)	Project support to groups important	Project approach evolving; reducing dependence on wider links
ASS, Ethiopia	Volunteer farmers' associations	To date, growers sell seed to project for use as source seed	Operations still controlled by project	Project hopes to withdraw, leaving self-sustaining local population
MIND, The Philippines	Committee members on project farm	Members take seed free; other farmers take seed and return after harvest (400 families)	Operations directed by project; local committees involved	High project involvement continues

The typical project profile is of a relatively large-scale operation, contrary to the common perception of local level seed projects as being very localised and small-scale. In The Gambia and Nepal, the projects are in fact replacing formal sector seed activity in the areas where they operate. The evidence from Mozambique suggests that GPSN could do the same, were the institutional barriers to this to be removed.

Organisation of Service Delivery

Improved timeliness of seed availability and easier physical access to seed should be major advantages of strengthening local seed systems.

A total of six different seed production systems are used by the agencies studied. The four most common are: contracting individual seed producers by the project; contracting community groups by the project; systems where community groups organise the production of seed themselves; and relying on the return of in-kind seed loans (used by the seed banks). There does not seem to be any connection between the type of agency, timescale of the project, quantity of seed produced and the production system used.

There are also a number of different seed distribution systems. Just under one third of the projects distribute seed via a mixture of direct sales by the project and sales by growers; relying entirely on distribution by growers, cash-based or otherwise, and using the seed bank system are equally common. Some use a two-stage process involving initial distribution to farmer members and subsequent distribution by these members. Only MCC relies entirely on private traders, although FAO and KHAP are experimenting with this.

From the evidence available, improved timeliness of seed availability is by no means guaranteed. Most of the relief-based activities, because they have involved shipping in and distributing seed from outside the area, are highly dependent for success in this regard on the availability of finance, transport and seed. The projects supported by OXFAM, Concern and CIC have all experienced problems caused by delays in the release of funds by donors.

Projects like the Sahel seed banks, MIND, ActionAid, CESA, SCF and MCC that work through grassroots membership organisations, have some idea of the number of households reached directly (see Box 7.1). But none of the agencies have investigated traditional community seed distribution mechanisms in any detail and few have attempted to trace how far project seed has spread and to which social groups. Comments made by agency staff and the limited number of articles on this subject (see for example Green, 1987; Cromwell, 1990; Sperling and Loevinsohn, 1992) suggest that the community diffusion of seeds may be more discriminatory than generally assumed.

Most of the agencies assume that seed transactions will be cash-based. Some of the projects have modified their charging system for seed sales, either moving from a less to a more formal system (Concern and OXFAM) or moving to less direct project involvement in sales (AA–N). A number maintain more than one system, involving preferential access to seed for growers plus cash sales to other farmers. Nearly a quarter of the agencies distribute seed as in-kind loans to be repaid after harvest. This leaves the projects at the mercy of the harvest; when the harvest is poor, seeds given out are not repaid. Attempts to set up village-operated seed banks in marginal, variable environments have repeatedly foundered on this non-replenishment of stocks after harvest failures.

A quarter of the agencies leave the method of distribution up to the growers; this may involve seed gifts and in-kind loans as well as cash sales. However, the majority distribute all seed as cash sales, either by the growers or by the agencies. This has potentially severe implications for the social impact of the projects because all the evidence shows that a significant minority of most farming communities do not have the resources to pay cash for seed (see above, pp. 30-32). Supporting local level seed production would therefore appear to achieve little in terms of wider access to seed unless it includes non-cash methods of seed distribution.

Internal Organisation and Operational Efficiency

Planning and Implementation
Most of the agencies surveyed have the necessary internal administrative systems in place to be able to plan and implement their work effectively. They achieve significant results in terms of output and farmer contact. However, a number achieve this at the cost of having to absorb substantial overhead expenses. In part this is because they tend to learn by experience as a substitute for developing effective linkages with organisations that could provide relevant advice or support. Related to this, there is some evidence of duplication and even competition between the services the agencies provide.

Monitoring and Evaluation
It is not always clear that the agencies have a good grasp of what their programmes are achieving. In particular, routine record-keeping and formal monitoring are not always in place. Most of the periodic evaluations carried out seem to have been requested by external donors. In some cases, seeds work is a new activity for the agency. Often the outputs of the agencies' other activities are easily visible—digging wells, building schools, etc.—and monitoring can consist of little more than physical verification of the work done. In these situations, seeds tend to be treated the same way and monitoring focuses on the number of hectares under seed multiplication and perhaps, but not always, the quantity of seed harvested. Questions about the quality of seed and its impact on local farming systems, and about its use and distribution, are often not addressed systematically.

Technical Services
Technical expertise in variety evaluation and seed production management varies widely between the agencies.

Some of the projects are not run by agriculturalists, and within these there appears to be ignorance about the importance of even basic technical aspects of seed care (isolation distances, the effect of temperature and humidity on the viability of seed in store, etc.). Some of the agencies devote little attention to providing agronomic and other technical advice to farmers within the project, either themselves or by outside agencies. For example, in most of the relief operations, it is left to farmers to deal with the seed they receive as best they see fit. In some cases this is not problematic; however, farmers are not always expert in seed use. For example, some of the members of the ACORD seed bank were refugees who had relatively little previous experience of crop cultivation.

At the other extreme, some agencies organise relatively large-scale or centralised seed multiplication. They have ample capacity, in terms of trained staff and equipment, to inspect and test seed as well as to participate directly in seed treatment, storage, etc.

Most projects lie between these two extremes. They are aware of the technicalities of seed production and seed care and provide assistance themselves, or facilitate liaison with government services. This is an important part of the work of a number of the agencies in The Gambia and in Nepal, for example.

Few of the projects studied had surveyed the project area in advance to identify the appropriate range of technical seed services to be supported. This is particularly obvious in relation to the relative weight given to agricultural research (local testing of modern varieties, evaluation of farmers' varieties, selection and other genetic improvement techniques) as compared to seed production, and the relative balance between seed production and seed distribution activities. It appears that most of the agencies make these decisions primarily based on their own policy agenda.

Project Financing

Project seed costs and prices charged are given in Table 7.1. Only Concern has distributed seed free; efforts have been made subsequently to encourage the recipient communities to view the seed as in-kind loans for starting seed banks. Only CESA has made seed available at less than local market prices; this was a special price charged to growers in the first year of the project, to encourage uptake of potato seed produced on community plots. By far the most common system is for seed prices charged by projects to be around local market prices for seed, with some agencies adding on handling and wastage charges.

The relationship between project seed prices and those charged by government agencies varies from country to country. However, government seed prices are usually distorted in one way or the other, compared to local market prices, and this creates problems for the projects. In Mozambique and The Philippines, government seed prices are higher than local market prices

Table 7.1:   **Case studies: economic cost of seed**

Agency	Example crop	Project seed selling price	Local seed price	Government seed price	Costs absorbed	Estimated economic cost of seed
ACORD	Millet	Local price plus handling	–	–	Store construction, disinfectant, training	–
ACORD	Millet	0.08	0.10	–	$240 grant, training	–
OXFAM	Millet	Local price plus wastage	–	–	Cement for stores, initial seed stock, transport	2.82
CONCERN	Millet	Free	1.10	0.69	Transport	0.85
CIC	Maize	0.52	–	0.74	Technical assistance, equipment, some chemicals	0.85
CESA	Potatoes	Local price less 5%	–	–	None after first year	0.11
MCC	Soyabeans	0.30	–	–	Grower subsidy, transport, overheads	1.04
AA–TG	Maize	0.30	0.25	0.22	Grower subsidy, transport, handling, staff	1.02
SCF	Maize	0.38	0.38	0.22	Transport, overheads	1.63
Good Seed Mission	Maize	0.50	0.50	0.22	Overheads	1.83
FFHC	Maize	0.38	0.38	0.22	Technical assistance	–
FAO	Maize	0.25	0.38	0.22	Technical assistance to dealers	0.27
PPS	Rice	0.24	0.24	0.22	Technical assistance, seed bin subsidy	–
KHAP	Rice	0.24	0.24	0.22	As above plus certification, revolving fund	2.40
AA–N	Rice	0.26	0.24	0.22	Technical assistance, seed bin subsidy	0.31
ASS	Various	Local prices	–	–	Technical assistance, transport, storage	3.24
MIND	Rice	Local prices	Less than govt. price	0.39	Capital costs, training	More than govt. price

Source: Case studies
Notes: Prices = US$ per kg 1990–91; – = not available; Economic cost calculated from annual project budgets and seed production.

for seed. This means that, where the projects have to abide by prices set by the government they are unable to compete with traders offering 'seed' in local markets. More commonly, as in Sudan, The Gambia and Nepal, government seed prices are lower than local market prices; none of the agencies reviewed can produce at these lower prices and so, again, the projects face stiff competition, at least in theory. In practice, the government service may be very limited and so not a threat. In Nepal, the government recognises this difficulty and removes government supplies from some of the areas where local seed supply projects are operating. CIC/GPSN and some of the relief-oriented agencies have faced competition from free seed provided under the emergency relief programmes of other donors. Charging local market prices for seed by no means recovers the full economic cost of producing it. Economic costs range from 20 per cent more than the price charged (AA–N) to 10 times more (KHAP), typically being about 3.5 times more—although projects vary considerably in what items they treat as attributable costs. Staffing and administration overheads are usually absorbed elsewhere in the agencies' budgets; some agencies also absorb the capital and other start-up costs (such as training) associated with their seeds work. Some agencies absorb the costs of transport, which can be a major item in relief operations and in projects dependent on seed produced by growers being taken away for testing, storage, and sale. A quarter of the projects also provide other subsidies: the cost of seed bins in Nepal are subsidised; and MCC and also AATG have subsidised the price paid to growers for seed. Only CESA no longer absorbs any project costs: these remain considerable and the peasant federations involved remain dependent on external grants to cover them.

Comparisons between projects within Sudan, The Gambia and Nepal suggest that the economic cost of the seed produced does vary between different types of agency. In The Gambia, seed costs are remarkably similar for the three NGO agencies for which there is data (note that AATG's lower costs exclude staff costs). In Nepal, the costs of the donor-funded project (KHAP) are considerably higher than those of the NGO (AA–N), no doubt because KHAP has a high expatriate technical assistance input and is providing US$ 1,700 revolving funds each for some 12 co-operatives.

Externally funded organisations, which are usually independent of local revenues for continued activity, can often sustain these high costs, at least in the short-run, and the agencies tend to play down the significance of their high overhead costs. In many cases the agencies are carrying out other programmes within the community as well, so it is difficult to distinguish between seeds costs and those for the other activities. In addition, where projects include an element of community empowerment (see below), the costs of this can be substantial and not justifiably attributed to the seeds activity itself. Many agencies justify project costs either on the grounds of emergency need, in the case of relief efforts, or on the grounds that they are operating a pilot project which can demonstrate a widely replicable model and justify early investments.

However, few agencies have carried out detailed surveys to assess just how valuable their seeds activities are for local farmers. The demonstration effect can be important but this is difficult to assess accurately and many villages have already had some contact with formal extension services. The instruction in seed husbandry given to project seed growers is almost certainly beneficial to farmers when saving their own seeds. In the KHAP programme, for example, this advice, and the improved access to complementary inputs, is valued more highly by the seed growers surveyed than the income from seed multiplication, which is small. The requirement for seed to be paid for in cash on most projects (see above, p. 86) will preclude a proportion of the community from benefitting. Overall, it is too early in the life of most of the projects to judge whether the seeds work has brought with it these other benefits.

The 'demonstration effect' argument often leads to agencies declaring that their involvement in seed activities is temporary, and that sooner or later the local community or some other institution will take over the project. As discussed later in this chapter, the feasibility of this is frequently not clear.

It is also not clear whether scaling-up activities will reduce unit costs, as a number of the agencies argue. It is true that technical assistance staff costs account for high proportions of project overheads in many cases, and these may be expected to lessen over time. However, it is much less certain that other running costs such as field inspections, processing chemicals, packaging costs and transport will be reduced. Like many crop and animal goods produced by biological processes, there are only small economies of scale in seed growing and a range of technologies can be used in production. Hence small farms can compete in seed production in a way that small workshops could not compete with tractor or fertiliser factories.

Community Empowerment

The notion of 'empowerment', from the ideas of Paulo Freire, the Brazilian educationalist, refers to the creation of an environment of enquiry in which people question and resist the structural reasons for their poverty, through learning and action. Many of the seed projects are closely associated with attempts to empower local communities; in the belief that the adoption of innovations cannot take place unless the capacity of the receivers is properly developed. Seeds are seen as part of a strategy for helping farmers to restore control over the technology they use, by offering alternatives to the standard packages of chemical inputs provided by formal sector agencies. In all cases, it is not yet evident to what extent such aims (which are intrinsically difficult to assess) have been met.

In some of the other projects, the emphasis is on agricultural improvement as a route to greater household welfare. One of the longest operating programmes, MCC, combines careful targeting of poor farmers in a risky environment with the promotion of imported, hybrid winter vegetable seed. The disadvantages of dependence on exotic seeds appear outweighed by the chance for farmers to produce more from the land they cultivate.

Community Participation

All but a minority of projects have mechanisms for community participation. Donor-funded projects have tended to opt for greater agency control; while the relief operations and ActionAid and SCF have promoted control by village groups. There is a correlation between agency control and technical difficulty and scale of operations: where the operations are both simple and restricted to the ambit of the village, local village committees are in control; where the project is more complex and involves more widespread cooperation—for example, soya seed production—the agency is in direct control.

Two thirds of the projects involve some degree of community control. Apart from Concern and CIC, which have both used existing local government structures, the rest have all introduced new project-specific village groups. For a number of agencies these groups have not apparently made substantial moves towards achieving real local control of seed activities and sustainable local capacity in organisation and decision-making: relatively high agency inputs into planning, implementation and supervision have continued. Also, there has often been little participatory needs-identification before the start of the project—although there are exceptions.

Box 7.2: Criteria for identifying farmer seed growers

Most projects agree that it is the more 'progressive' farmers who are most likely to be successful as seed growers and there is a set of characteristics of the ideal farmer-grower which appears to be universally accepted. Farmer-growers need to:

- have relatively large holdings, so that they can allocate land to seed production with jeopardising domestic food production;
- be well-integrated into existing extension services, so that they get the advice necessary to grow seed well;
- be relatively well educated, so that they are able to understand and follow the advice given;
- be relatively more commercialised, so that they can afford to buy the inputs necessary for seed production;
- have some kind of leadership role in the local community, to provide a demonstration effect to other farmers;
- have relatively good quality land, so that they can get the yields necessary to justify the extra inputs used in seed production.

Sources: Pinchinat in CIAT, 1982; PAC, 1986; Berg *et al.*, 1991.

Most of the projects surveyed aim to work with poorer communities. In some cases, especially in Mali and Sudan, the beneficiaries live in marginal drylands. However, only in a few cases do the agencies specifically target poorer farmers within their areas of operation; elsewhere, little attention is paid to differences within villages and households. Consequently, the relatively better-off and more powerful, with their extra resources and easier interaction with outsiders, are in most cases the first to benefit from the projects. Moreover, where projects are involved in multiplying as well as distributing seed, the farmer-multipliers have to be selected on the basis of their skill and resources, thereby favouring the better-endowed growers (see Box 7.2). The difficulty of allowing equitable involvement in local seed multiplication whilst ensuring technical standards (which requires a relatively high degree of centralised direction) has been documented by numerous seed projects and programmes in developing countries.

This highlights the difficulties of attempting to make a seed project provide both an income-generating activity for seed growers and a cheaper seed service for small farmer seed users. Most agencies seem to perceive their activities more in terms of one than the other. For example, most of the NGOs are aiming to provide almost a social service for the smaller, poorer farmers within their communities and are not concerned with the possibility of establishing private commercial seed growing; MCC, FAO and the projects in Nepal, on the other hand, explicitly aim to support an income-generating activity by helping small farmers to capture some of the benefits of seed production.

Many of the agencies intend seed to be distributed through existing community mechanisms, assuming that by this means the benefits will soon spread throughout the community. However, in no case has this been verified by investigation (see p. 86).

Links with External Agencies

The extent to which the agencies have actively sought links with the other local and national institutions involved in seed supply has varied significantly. A number have tried to operate within the national seed sector policy framework from the earliest stages of planning and have perceived their role as supporting existing government institutions. A minority see their role as an alternative to any formal sector seed activities operating in the area. Most agencies are supporting local seed systems because there is no government service in the area in which they are operating.

In The Gambia and Nepal, it is government policy to work with NGOs and there has been genuine collaboration. Similarly, the relief operations in Mali and Sudan receive official blessing and support. Generally, the agencies form their main links with government bodies but they also work with private traders and with other NGOs. In some cases this is problematic: there have been particular problems with co-ordinating different agencies' seed activities for relief and rehabilitation. In others, however, networking between NGOs has been successful. For example, OFSP, the project for networking and technical support to NGOs active in seeds in The Gambia and Senegal, has been commended by both NGOs and government bodies. Similarly, MASIPAG (the coalition of farmers' organisations in The Philippines) has been an important source of farmers' rice varieties for NGO seed projects.

One important result emerging from this research is that there are problems co-ordinating with other seed institutions—particularly those operated by government—and with national seed policy, regardless of the degree of effort invested in trying to achieve this by the agencies involved. Where problems have been encountered, it is difficult to disentangle the problems caused by pressure on government budgets from those caused by antipathy towards the NGOs.

Sustainability of Seed Activities

The longest-running project is ACORD's seed bank programme in Mali, which has been operating for 15 years. However, this has undergone radical changes during its course and its long-run future is currently again in doubt due to drought and war. Most of the projects have not been operating for so long—typically for around five years. Many of these have also had to be rethink their activities at some time. FAO's approach in The Gambia is still evolving.

CESA and AA–N are making progress towards sustainability. Their projects have a common profile: community control or a strongly collaborative approach from the start; seed distribution through existing community mechanisms encouraged from the start; seed quality standards adapted to fit community needs and capacity; and a generally low external input seed production strategy. As a consequence of this, both projects have low overheads.

All the other agencies express the desire to hand over projects to community control to ensure long-run sustainability, but are experiencing problems in achieving this. For half of them, this is due to continued harvest failures threatening seed production or seed bank replenishment. For the other half it is due to problematic links between the project and the wider national economy, including problems with supplies of complementary inputs and problems with government extension and seed certification services. The cost of compensating for these failures is high and pushes the cost of project seed up.

Only CIC has supported a permanent and formal seed production agency. The other agencies generally see their role being taken over in the longer run by village groups or private farmers and traders. Only in the case of the Bangladesh soyabean seed multipliers are there clear indications that private actors might independently continue seed production. In other cases, hopes for small-scale private seed enterprise have been expressed, but it is not clear whether the seeds produced can be sold at a premium, and whether the producers can link up with supplies of foundation seed and services for testing and certification.

Local groups as seed producers have had mixed fortunes. Relatively simple operations like seed distribution and running seed banks, as in Mali and Sudan, present few problems to village groups, although they cannot overcome harvest failures. When it comes to more complex tasks, such as obtaining foundation seed or trying different seeds in trials, or testing seed for quality, no group has yet been able to assume responsibility.

Nevertheless, half the agencies are planning to withdraw after a specific time, and hand over operations to local groups. In some instances, agencies argue that since local groups can assume responsibilities for running schools or maintaining local roads, they can also multiply seeds—with little recognition of the difference in the range of skills required. Critical assessments of the feasibility of agencies' plans for handing over seed activities to local groups are rare. The failure to produce such assessments appears to be the result of agencies' lack of awareness of two crucial issues. First, the institutional role that they fill (replacing, supplementing or providing an alternative to existing local institutions) varies from place to place, according to local circumstances. What is the most suitable structure will vary dramatically depending on the type of activity (breeding, multiplication, distribution; modern varieties or farmers' varieties, etc.) and the existing institutional structure and farming system. Second, each of these roles means that the requirements for a successful project handover from agency to local institution will be different.

International Seed Advocacy

The pressure groups reviewed in Chapter 6 are involved in global lobbying on seed issues rather than in direct seed production and distribution so their performance has to be assessed using a modified set of performance indicators.

Both GRAIN and RAFI developed out of the ICDA seeds campaign and have thus had a relatively long history, albeit not in their present form. They both have a small, committed staff and low budgets and produce a large output of information. As well as publications, conferences and other tangible results, such as the various international agreements which they have lobbied for, they provide considerable indirect support for Southern groups both practically and via their campaigns.

They fulfil a useful function which is likely to become increasingly important as issues relating to plant genetic resources, such as patents and plant breeders' rights, assume greater international importance. Their importance will also increase as Southern groups, which often find it difficult to maintain international contacts, play a greater role in local development initiatives. GRAIN and RAFI have, as umbrella organisations, publicised the work of three of the projects reviewed in this book, CIC, ASS and MIND. Providing this kind of channel for feedback of projects' practical experiences is an important function for successful international lobbying.

However, they will have to maintain a fine line between providing the emotive, simplistic, single issue literature on genetic resources needed to influence general public opinion and providing the more balanced and technically comprehensive information needed to influence international organisations and to support practical projects in the South.

Advocacy groups are less effective in directly informing and influencing the seed project activities of the Northern donor-funded and NGO agencies. GRAIN and RAFI are aware of this and have made this a priority area for future work. Progress in this area will depend as much on the agencies' approach as on the value of the work carried out by the pressure groups. The majority of agencies reviewed here have relied on their own experience, rather than seeking to learn from the experience of others, at least at the project planning stage. Communication between project agencies and advocacy groups needs to be improved.

In addition, the larger NGOs should themselves consider whether they could engage in this kind of advocacy work, both towards Southern governments and towards policy makers in the South, without compromising their apolitical position. While GRAIN and RAFI play a catalytic role and have had some major successes, ultimately their role is limited by their size.

Typology of Support for Local Seed Systems

Based on the evidence presented so far, it is possible to distinguish a typology of agency involvement in seed activities, in order to compare the approach of different types agencies. Four main organisational types can be distinguished—North-based direct action NGOs, North-based NGOs supporting seed projects in the South, South-based NGOs, and donors funding government projects.

North-based direct action NGOs. One third of the agencies reviewed in this book are North-based NGOs (ACORD, OXFAM, Concern, MCC, SCF). Their seed projects have their origins in relief operations and most are aimed at community development as well as at increasing household production. Half of these projects entail substantial community involvement, mainly via seed banks, and substantial quantities of seed are provided.

All the projects appear to be operating relatively efficiently but the agencies absorb a significant proportion of the overhead cost. Seed, where it is sold for cash, is often sold at only 25 per cent of its full economic cost.

Half of these projects are currently threatened by harvest failure. Although the agencies plan to withdraw their support from the remainder in the long run, the continuing need for high staff inputs make this unlikely in the near future.

North-based NGOs supporting seed projects in the South. One third of the agencies reviewed are North-based NGOs supporting seed projects operated by local organisations (Action Aid, CIC, ASS, GRAIN, RAFI). All of these have as their main aim the empowerment of local communities as well as increasing household production. Two of them, GRAIN and RAFI, are not directly involved in project activities and instead fulfil an advocacy role in the North on behalf of projects in the South.

The degree of community involvement in the seed production and distribution systems used by the agencies is mixed: only half leave project organisation to community groups. Those projects which make seed available to the community at large do so via cash sales (half of them are new and therefore use seed produced as source seed for subsequent multiplication). The quantity of seed produced is in the middle range (7–185 tonnes) compared to the output of other types of agency.

All the agencies absorb the cost of project overheads, some of which are substantial, and seed is sold at between 25 and 50 per cent of its true economic cost of production. Many of the agencies in this category are taking active steps to reduce their involvement in the project, but it is proving difficult to ensure the necessary linkages between the projects and other external agencies.

South-based NGOs. Two of the agencies reviewed are South-based NGOs supporting local seed systems. Both aim to empower local communities, CESA by providing an alternative to government seed services, which are failing to reach small farm communities in the high Andes, and MIND by providing an alternative to government seed services in The Philippines which do not cater to the needs of small farmers. Both projects are collaborative ventures with local communities, one on project land and one on community seed plots. Relatively small quantities of seed are produced.

CESA's project seems to operate relatively efficiently and to have considerable potential for achieving long-term strengthening of local seed systems: it has not needed outside financial support since the first year. MIND's project continues to have a large agency involvement and there are unresolved questions about seed quality and about the long-run economic viability both of the project and of the use of local varieties by small farm households. Also, it has suffered from high staff turn-over and a number of project reorganisations.

Donors funding government projects. A quarter of the agencies reviewed are multi- and bi-lateral donors or technical support agencies, supporting local seed systems through government projects (FAO, PPS, KHAP). Their main aim seems to be to encourage existing government services operating at local level to use alternative service delivery approaches, in order to increase household food production, and in particular to increase the use of private traders. All these projects have overall control of seed production planning and all distribute seed via cash sales. All produce relatively large quantities of seed.

They appear to operate efficiently within the project approach that they have adopted, but they all have relatively large overheads. Seed is sold at as little as 10 per cent of the true economic cost of production.

All hope to withdraw from project support in the long-run but are experiencing difficulties in achieving this, due to difficulties in links with other external agencies.

The case studies thus confirm a number of common assumptions about the differences between the approaches of the different types of agencies. The donor-funded agencies tend to place less emphasis on community empowerment and more on increasing agricultural production compared to the NGOs. However, importance attached to aims of relief, development, empowerment and advocacy varies substantially between NGOs. This distinction is reflected in the strategies of the agencies. Donors place greater emphasis on increasing economic efficiency and growth through the use of modern varieties and better quality seed, compared to NGOs, which tend to emphasise reducing risk and dependence on external agencies, through diversification of varieties and increased use of farmers' varieties.

Having said that, there appears to be little correlation between the aims and strategies of the different types of agencies and their primary activities: the blend of agricultural research compared to seed production and seed distribution; the blend of local varieties and modern varieties; and the blend of services provided to seed growers and to users.

For only two of the agencies (PPS and ASS) is support for local seed systems their primary activity in the area in which they are operating. The majority have introduced this support as an adjunct to existing relief operations or rural development programmes. Only half the agencies have employed agriculturalists or specialist seed staff to implement their seeds work.

Only MIND has provided support for local seed systems as an alternative to a government system which is functioning but felt to be inappropriate to the needs of small farmers. And only one (ACORD in Mali) has provided support in a way which depends on government providing some input (the co-operatives). The vast majority of the agencies are supporting local seed systems because no government service is reaching the area in which they are operating.

Does the NGO approach to development provides an alternative development model for strengthening local seed systems? NGOs have demonstrated that seed improved either genetically or physiologically can be produced at the local level by decentralised systems of production. Nevertheless, these achievements are still small-scale, costly, and not without defects. The most obvious conclusion is that there is a dearth of projects that are compatible with strengthening community control of seed systems (as represented by the blank boxes in Diagram 2.1). Few NGOs are using existing community structures and working with local varieties or adapted modern varieties appropriate to small farmers' needs. These appear to be the South-based NGOs. Most instead set up new local seed multiplication and distribution systems and work with modern varieties produced by formal sector agricultural research. This has serious implications for the long-run sustainability of the local seed systems supported by NGOs, which over-ride NGOs' expressed aims of empowerment, community control and responding to felt needs.

8
Conclusions

The Case for Supporting Local Seed Systems

The evidence presented in this book suggests that there is considerable development value in supporting local seed systems in four situations:

- in communities *where there is no seed at all,* any seed activity will be well received by farmers;
- where support *increases the benefits small farmers derive from the farming system.* This has various dimensions: it could entail the introduction of new varieties, better quality seed or a new crop. But care has to be taken to ensure that this kind of change does not discriminate against less well-resourced farmers by, for example, requiring additional external inputs such as irrigation or fertiliser;
- *in a defensive capacity,* where the formal seed sector does not provide varieties suitable for small farmers. Here the need is often simply for clean seed of farmers' existing varieties, which is not otherwise available, either because the formal sector does not provide it or because farmers cannot afford the available supplies;
- where the continued use of *current seed sources will increase risks or reduce yields.* This is the case where, for example, declining and increasingly variable rainfall means farmers' varieties are no longer well-adapted to the local environment; where farmers' varieties are genetically degenerated or prone to pests and diseases; or where modern varieties are becoming increasingly susceptible to disease.

The corollary to this is that in other situations, external support for seed systems may be of little value because it is inappropriate to the farming system. This has been the case, for example, with the modern millet varieties distributed in the hill zones of Nepal; and with the medium-duration sorghum introduced by relief agencies in Sudan.

This categorisation distinguishes different physical environments. This is only the first cycle of questions that have to be asked to understand the potential for supporting local seed systems. The second cycle is as important and relates to the wider socio-economic perspective: whether better local seed systems are what communities can best use in order to improve their livelihoods. This wider question is too often ignored. Few farmers anywhere in the world, in the North as well as in the South, are full-time farmers: surveys show consistently that an average of 40 per cent of household income is derived from off-farm activity, because returns to agriculture are often lower than returns to manufacturing or service activity (Kohl, 1991; Low, 1986).

It is not only the proportion of income earned off-farm which is important, but also the proportion of resources which households devote to off-farm activity. Some households cannot make use of new varieties because they do not have the extra inputs necessary to produce a worthwhile extra yield from them, or because returns to the investment of these resources in other areas are higher. However, new varieties can be highly appropriate for households who are seeking not to maximise but to guarantee production using minimal labour inputs. In this case, the use of hybrid varieties and fertiliser can secure domestic food supplies with lower labour inputs than traditional local varieties. This has been documented by Low (1986) for women maize farmers in Swaziland, for example. The most appropriate mix of varieties depends on the institutional and agricultural research context and it is not necessarily helpful to promote reliance on local varieties and the preservation of a high degree of genetic diversity in all circumstances. This point is discussed below (see p. 112).

It is therefore critically important to understand the context of local seed systems and to identify needs accurately before planning support strategies.

Key Areas of Support for Local Seed Systems

The experiences documented in this book have contributed to our understanding of the key areas in which support for local seed systems can be beneficial.

Varieties. Farmers may want access to seed of a new crop, to seed of new varieties of a crop that they already grow or to fresh seed of varieties already in use. Which is required will vary both between communities and within them. Both farmers' varieties and modern varieties, particularly those based on selections from local landrace material, may be needed.

In general terms, it is usually most helpful to provide a diverse range of varieties, and also material with substantial intra-varietal variation. It is also usually beneficial to concentrate on material which has a low requirement for external inputs and which meets farmers' needs in terms of non-yield characteristics such as taste, storage or straw. Farmers are most likely to try out new seeds for cash crops planted in pure stand, for open-pollinated crops with low sowing rates, and for crops which are prone to seed-borne diseases or to rapid deterioration in store.

Seed quality. Farmers are often adept at maintaining physical seed quality and may not require specific support in this area. However, there may be specific problems in certain areas, in which case a substantial increase in quality can often be produced with relatively simple technical innovations (see above, p. 28).

This does not necessarily mean the use of ISTA standards, though, and alternative safeguards for purity, germination, etc. may be sufficient. Assistance with pest and disease identification and with simple improved storage techniques and technologies are often among the main requirements.

Whatever the context, external agencies cannot ignore seed quality issues.

Seed diffusion mechanisms. The most appropriate organisation of support for local seed systems in terms of ensuring long-run sustainability of seed activities can only be established from a thorough and participatory initial needs identification.

Using existing community structures for seed diffusion, rather than setting up new village groups, is often more likely to ensure the sustainability of the initiative in the long run. However, some modifications may be needed to ensure that the poorest groups are able to participate.

For many initiatives, especially those that are designed to empower local communities to interact with external institutions, agency support may be needed for a relatively long period.

Linkages. Linkages with external seed sector institutions supplying extension services, complementary inputs, seed certification services and plant breeding expertise are essential. It is possible to minimise the need for such linkages by using low input seed production and distribution systems, but they will still be needed to some extent and they have a major influence on the long-run sustainability of local seed systems. It is important to provide support for strengthening such linkages.

One of the most important linkages is between national agricultural research services and the farmers they seek to serve, as this has a critical influence on technical change in agriculture. This is discussed below, p. 116.

The Context of Local Seed Systems

What needs to be done to support these key areas depends on the context in which the local seed system is operating. Four factors in particular are important—crops, climate and agro-ecosystem, local community structure, and national policy:

Crops. Different crops have different technical requirements, as we saw in Chapter 2 and the appropriate blend of plant genetic resources varies between farming systems. Effective support will therefore vary according to prevailing cropping patterns.

Climate and agro-ecosystem. It is difficult to create sustainable seed systems in the fragile agro-ecological environments in which many small farmers in developing countries live. In dryland areas, each season is different and a high proportion are failures. Agricultural projects in these areas often fulfil a kind of 'insurance' function, and perhaps 40 per cent of project costs are, in essence, insurance 'premiums' rather than investments in sustainable seed supply.

Recognising this is important when agencies are assessing their long-term role in marginal, variable areas, where there are unlikely to be either suitable varieties developed by the formal sector or the institutional structure necessary for agencies to work synergistically with formal sector institutions. It appears that many of the seed bank initiatives in the Sahelian zone have fulfilled this role, regardless of the initial intentions of the agencies supporting them.

Local community structure.   Existing local community structures have an important influence. The nature of that influence will be affected both by the degree of social differentiation within communities and the extent of their links with external agencies.

Better-off members of the community may be able to appropriate new seed production knowledge and to preclude poorer and less powerful groups from participating in supported seed systems. Existing seed diffusion mechanisms, in particular, may not be equitable. Access may be limited to certain ethnic or social groups or access mechanisms may perpetuate poverty through, for example, requiring large quantities of seed to be returned in payment for in-kind seed loans. In other areas, structures may be very equitable. For example, many Muslim communities in Sudan and Mali operate a seed tithe which is planted out in community seed plots (Renton, 1988), which has strong potential for developing as a supported seed system.

Regarding external links, a recent report on the decentralisation of renewable natural resource management in the Sahel (ARD, 1991) provides a synthesis of the basic requirements for successful community resource management which is highly relevant to community management of seed systems. The report suggests that communities need to be able to undertake collective action; to facilitate private sector activities; to co-ordinate initiatives for the local management and governance of resources; and to solve conflicts. Then they will be able to create and sustain institutions for the local management and control of seed activities that can mobilise and manage labour, equipment and funds and that are willing and able to work with external agencies.

National policy.   There are many dimensions of national policy which influence local seed systems: seed quality control standards and other legislation affecting the seed sector; controls on the types of institutions allowed to operate in the national economy; and controls on the type of services that formal sector institutions are allowed to provide (for example, the degree to which agricultural researchers can work with NGOs and how far they are oriented towards small farmers). One of the most important factors is the state of the market, including market information systems and pricing policies. It is not uncommon for there to be little prospect of selling local farmers' varieties in the official market because crop authorities buy only modern varieties. In this case farmers' varieties are not attractive to farmers as cash crops, regardless of how well they yield. Ways in which policy can be tailored to the needs of local seed systems are discussed above (see p. 116).

Lessons for Improved NGO Support for Local Seed Systems

Whilst these contextual factors exert a critical influence on the success of support for local seed systems, the evidence in this book has shown that the internal organisation and operation of agencies' programmes is of equal importance. Organising for internal efficiency and for long-run sustainability are equally important.

Organising for internal efficiency

Administration and programme planning. Administration and programme planning must be organised to minimise costs and maximise effectiveness. Three factors contribute to this. First, detailed advanced planning is required. In the KHAP project in Nepal, for example, the number of seed producer groups that could be supported effectively was initially over-estimated by the project, with the result that individual groups developed as cohesive units more slowly than hoped. Box 8.1 outlines the key administrative aspects of support for local seed systems that require advance planning.

Second, an important lesson from the case studies is that the successful agencies all know the area in which they are working very well. And third, agency staff work well as a team and the agencies allow individual staff, who are often highly committed to the project activities, to shine.

Technical expertise. Strengthening local seed systems frequently involves providing technical as well as institution-building support. Agencies do not always appear to recognise the importance of technical expertise. Ensuring that an appropriate set of techniques and technologies is chosen for on-farm seed production is critically important. These must cover production, harvesting, drying, cleaning and, in particular, storage. Ordinary storage is often insufficient in hot, humid tropical conditions.

Success depends on obtaining a clear idea of the level of indigenous technical knowledge before the project support starts and on project staff being able to make sound technical recommendations for improvement. Although in general NGOs need to build on existing community seed maintenance and diffusion systems, this does not necessarily mean that no technical changes are needed. However, putting in place new production methods that replicate the formal sector high-input mechanised model can endanger the long-term sustainability of local seed systems. Examples of the kinds of techniques and technologies that can be useful have already been given above (see p. 28).

Charges for seed. Producing seed is more expensive than producing grain,

---

Box 8.1: Advance planning for local seed system support strategies

The following aspects of support for local seed systems need thorough advance planning:

- identification of farmer seed producers;
- source of working capital for farmer producers;
- type of material to work with (farmers' varieties, modern varieties, blend);
- source of foundation (source) seed;
- seed quality standards;
- nature of technical assistance and manner in which it will be provided;
- seed processing and storage equipment;
- market strategy (publicity, pricing policy);
- budget;
- phasing (pilot phase to evaluate varieties and how best to organise production and distribution);
- strategy for ensuring long-run sustainability (technical self-sufficiency and local control).

---

regardless of the production system used. At the very least, extra labour is required for roguing the crop and for sorting usable seed from rejected material after harvest. Thus, under free market conditions of supply and demand, seed must fetch a higher price than grain for any to be produced. Minimum grain:seed price differentials are given in Box 8.2.

This means that, on the demand side, there must be a real demand for seed from farmers: there must be a role for better planting material both in the physical conditions under which they are farming and in the economic context of their farming activities. On the supply side, the seed provided must have the potential to generate a tangible increase in productivity (or a reduction in the riskiness of production), either through its superior genetic potential or through its better physiological quality.

There are therefore various conditions under which there will not be a premium price for seed. First, if the seed is poor quality, either genetically or physiologically. Second, if the environment is harsh and using better seed will have little effect. Third, if farmers are not seeking to increase on-farm productivity—for example, because the opportunities for earning income off-farm are better. Fourth, if the market for seed is not determined by the competitive forces of demand and supply—for example, where official seed prices are subsidised or where some seed is distributed free by donors or other agencies. (Selling prices set for farmer products, which are also commonly controlled, also have an important influence on the market for seed.)

NGOs therefore have to work through a check-list of questions in order to arrive at an appropriate charge for the seed produced by the projects they

---

Box 8.2: Minimum grain:seed price ratios for different crops

Crop	*Ratio*
Single cross maize hybrid	1:5
Three way cross maize hybrid	1:3
Double cross maize hybrid	1:2
Groundnuts	1:2
Wheat	1:2
Rice	1:2

Note: Factory gate cost (i.e. processed and packed); ratios in developed countries are often higher.

Source: Cromwell, Friis-Hansen and Turner, 1992.

---

support.

First, agencies must establish, through the kind of initial needs identification outlined in Appendix 2, that access to better seed will be useful to farmers in their current farming system and socio-economic system. Second, they must be certain that the kind of seed to be made available (whether it is a better quality farmers' variety, or new genetic material) is the right one for the local farming system.

Third, they must ensure that the seed production system used minimises the cost of the seed produced. Some agencies introduce seed production and distribution systems which borrow unnecessarily heavily from the formal organisational structures and quality standards of large-scale national seed projects and programmes, thereby incurring high costs and making it difficult to achieve economic viability in the long run. The cost of supporting local seed systems can be much lower than the cost of operating national formal sector systems.

Fourth, NGOs should make every effort to ensure that demand for the seed produced is as strong as possible. This may involve on-farm demonstrations of the performance of the new seed and also extension work to explain the seed care and crop husbandry techniques necessary for the new seed to be used to best effect.

All these avenues should be explored fully before any arrangements are made to subsidise the price of seed produced by the project. If, however, all these components of the seed system have been assessed correctly and the price of seed made available through the project *still* results in a seed:grain price differential substantially above the norm, then NGOs are correct to consider subsidising seed prices.

There are three common situations where this may be the case. First, where the agency is trying to promote the use of local varieties but is facing competition from modern varieties provided at subsidised prices by government or other agencies. Second, where government controls crop product prices so that paying the relatively higher price of project seed is not economic for farmers. And third, where there is as yet no market at all for crops produced using farmers' varieties (for example, where these crops are traditionally used for domestic consumption only or where state marketing authorities buy only crops produced from modern varieties). In this situation, NGOs may decide that the wider long-term damage caused by the widespread use of modern varieties (the loss of local germplasm, the increased susceptibility to pests and diseases, etc.) may be great enough to warrant subsidising the cost of farmers' varieties seed, to encourage its use by individual farmers in the short-run.

In addition the provision of services such as seed production and distribution is always more costly in marginal areas than in the mainstream high potential agricultural areas of the developing world. This is because small farm areas are usually relatively remote from market centres, the population is scattered, individual farmers require only relatively small quantities of inputs and the terrain over which service delivery agents have to move is often difficult. Again, NGOs may consider that the importance of providing seed services to small farmers in these areas justifies a subsidy.

Various conclusions can be drawn about the prices that are charged for seed produced in systems supported by NGOs. Under normal market conditions, prices will never be as low as prevailing grain prices. Government price policy for seed and for crop products may distort seed:grain price differentials further. But there are nonetheless various steps NGOs should take to ensure that seed is produced as efficiently as possible and that it is a product for which there is a real demand. The most appropriate steps in any given situation require careful preliminary investigation: there are no hard and fast rules. In a number of situations (particularly where the project is producing seed of farmers' varieties), long-term subsidies for project-produced seed will be necessary. This demonstrates the importance of taking a long-term perspective in deciding the duration of agency involvement in a project and the level of funding required.

Linkages with external institutions. Forming effective linkages with external institutions, including other NGOs, is important so that agencies are not duplicating other agencies' effort or adding unnecessarily to their own costs. Not all agencies have been very good at doing this (see p. 118).

Organising for sustainability

Needs identification survey. Probably the single most important requirement for sustainability is for agencies to carry out an initial seed needs identification survey. They must be able to do this in a way which is comprehensive, detailed and, above all, accurate and which examines both the kind of material farmers want and the blend of technical and institution-building support that is needed. Appendix 2 lists the kinds of questions that a needs identification survey needs to answer before detailed planning of support can start.

Participation. We have seen the considerable range of community structures that agencies can work through in order to support local seed systems. Identifying the most appropriate structure to optimise participation is one of the critical requirements for the long-run sustainability of any initiative, but it is also an area in which there can be some very difficult trade-offs. The basic difficulty is that the best structures for an agency to work through are those that already exist within the community—but these have by definition been formed for another purpose. They are either indigenous community structures, about whose operations the agency may know little; or they are groups set up by other agencies, or by the same agency, for another purpose. The success of support for local seed systems is highly dependent on two aspects of the groups through which the agency works:

- **equality**: the groups will be subject to any existing biases within the local community;
- **administrative capacity**: indigenous community groups in particular may never have needed formal accounting and record-keeping skills before, but agencies require them to account for money and seed.

This calls into question the benefits of working with community groups: while working with such groups is one way of empowering communities, it can also become another means by which community elites obtain control of external resources. In the past, the heterogeneity of small farm communities has been overlooked by some agencies, on the incorrect assumption that working directly with a minority of the community will empower the majority.

Supplementary initiatives are therefore likely to be needed in three areas. First, alternative systems are needed for ensuring that seed reaches social groups without access to traditional diffusion mechanisms such as certain ethnic groups, women farmers or poorer households. For example, a limited quantity of seed may be targeted on these groups by being distributed through other local development agencies working with them, such as church groups or health projects. Second, if new varieties are to be introduced, special systems will be needed for channelling initial supplies of this seed into the community. Wherever possible, these should use existing channels, for example local markets, or alternatively the key seed diffusers within the community. Third, these key seed diffusers have an important role in the regular movement of seed around the community (see above, p. 30). To ensure that community mechanisms work as effectively as possible, it may be worth expending some effort to identify who they are and then informing other farmers in the locality about them.

One important observation based on experience in Latin America, where alternative forms of organisation for local seed systems have been taken furthest so far, is that the sources of support for the community seed system must remain outside the group. This strengthens the group by requiring it to seek outside help itself, when necessary (Lewerez and Poey in CIAT, 1982). The extent to which communities can take effective control of supported seed systems also depends to a significant degree on the operating cost of the system, which is itself affected by the way the system is organised. Elements to avoid, therefore, include: paying large premiums to contract seed growers (which community elites who seek to control seed production may press for); engaging in high cost seed processing activities, such as packaging, which are unlikely to be necessary in farmer-managed systems; and transporting seed over long distances as this adds dramatically to the total cost and leads to dependence on access to vehicles, fuel and spare parts. Despite the evidence in favour of using existing community groups as much as possible, many NGOs still seem to try to set up new groups for seed activities.

While group organisation is important, there is still much debate about whether seed production itself is best carried out by individuals or on a community basis. Some development workers insist that production has to be carried out by individuals (Bal and Douglas, 1992) but one of the most successful projects reviewed in this book (CESA) is based around production on community plots. The answer probably depends almost entirely on the traditional production systems used by the community.

Whatever structure is chosen, it is clear that large amounts of time need to be spent sensitising communities to the idea of support for their seed systems and preparing them for implementation. The inadequate amount of time devoted to this is one of the most commonly cited self-criticisms of NGO seed projects in subsequent reports and evaluations.

On the positive side, a number of agencies have found that the effort put into developing sustainable local groups for seed production has benefited other aspects of community development. The groups have become forums for the community to articulate wider development problems (as in the case of the NEF/OXFAM village seed banks in Mali) or channels for a wider range of new technologies and other innovations (as in the case of ACORD's seed banks in Sudan and KHAP's seed producer groups in Nepal).

Trade-offs. Agencies must be sensitive to diversity within small farm farming systems. This is essential if they are to cope with the different needs of groups of farmers within a given community and the different crops within local farming systems. It is important to recognise the compromises that have to be made in the face of conflicting interests. In the NEF/OXFAM seed banks in Mali, for example, seed is distributed equally amongst all villagers rather than on the basis of need, in order to ensure that the seed bank is seen as a communal enterprise in which everyone in the village has a stake. This is seen to be critically important for the project's success (NEF, 1988).

While introduced seed distribution systems may be more equitable, they may be less sustainable. One important reason for this is that cash-based seed sales, necessary for long-run viability, exclude the poorest households. Agencies may need to accept a compromise. One possible solution is to encourage diversified or multi-institutional seed supply systems: for example, a blend of farmer-to-farmer distribution, local market sales, project distribution, etc.

One of the areas where trade-offs are most obvious is in the selection of farmer seed growers, as was discussed above, (p. 91). This is because, to be effective seed producers, farmers usually need above average access to land and other resources. Projects may not be able to maximise benefits for both growers *and* users because ensuring a good return to seed production often involves setting retail seed prices relatively high, and vice versa (see p. 88). Agencies therefore need to identify which is the priority target group and set activities accordingly.

Finally, it is important to monitor the impact of introducing new seed selection and storage technologies from outside. In a number of cases, this has served to remove from women their traditional role as seed-keepers, thus jeopardising their access to an important means of production. For example, ensuring the participation of women in the local seed supply initiatives in Nepal has been a recurring problem (see above, p. 64). There have been similar experiences in Malawi, where the majority of farmers taking part in the Smallholder Seed Multiplication Scheme have been men, even though traditionally it is women who store and select for planting most kinds of seeds (Cromwell and Zambezi, 1992).

Phasing of support. NGOs need to recognise that the nature and duration of their role depends not only on external factors but also on how effectively they carry out their work. Some agencies seem too keen to move into an area with a pre-judged set of activities, based on their own ideology rather than on the felt needs of the community, and to ascribe all subsequent problems to external factors, ignoring the role of their own weaknesses and mistakes.

One of the most important requirements in this respect is for agencies to have a realistic strategy for sustainability, based on an accurate perception of their own institutional role over time. Phasing of support is important in order to reduce the risk of initial failure. The consensus of opinion is that it is unrealistic to try to introduce new organisational structures and new technologies at the same time. Structures should precede technologies and there has to be a flexible and evolutionary approach to project development that takes account of farmers' needs and potentials changing over time.

In seed supply NGOs are often required to act as substitutes: either as a supply-side substitute, for government or the private sector, or as a demand-side substitute, for local community structures that are unable to deal effectively with formal institutions (or as some combination of both). Because NGOs are fulfilling a 'missing' function they cannot simply withdraw from the community at the end of the project life unless another institution has accepted the capacity to take over their role. From the evidence accumulated for this book, it seems that a significant proportion of NGOs are not willing to accept this reality and persist with plans to withdraw from communities after providing relatively short-term inputs.

There are examples of communities in Latin America which have successfully banded together to buy in expertise, to influence policy and to demand services in other ways. The interaction of farmers associations in Bolivia with national seed services is documented in CIAT (1991), for example. Thus, for example, CESA in Ecuador can realistically consider reducing its role in its seed project in the medium term, because federated farmers' associations and a wide range of other institutional linkages exist which could take over its functions.

In contrast, in sub-Saharan Africa, small farmer organisation tends to be weaker and advocates on behalf of the dispersed rural poor are needed. The experience in The Gambia is a good example. The NGOs there are performing services that are critical for a strengthened local seed system: they are a substitute for government and private sector seed services in the poorer small farm areas. Handing over the projects in the medium term future is probably an unrealistic aim: the village groups that exist are isolated and semi-formal and could not handle these functions.

Long-term support is also likely to be needed in marginal, variable environments. The case studies have shown that seed banks in these areas cannot operate without periodic external support.

Thus NGOs have to be clear about their role in seeds activity. Are they working at a temporary stage in a dynamic process of community development, or they are working as advocates on behalf of powerless communities and therefore needed over the long term? Whatever the case, it seems that a much longer time-span is needed than many NGOs currently anticipate. The most appropriate time-span varies, but local level seed supply systems may take at least ten years to develop to a degree where they are sustainable without outside support, even in favourable circumstances (Verbrugt in CIAT, 1982; Gaifami, 1991).

Genetic Resources Policy

The need to maintain biodiversity and the dangers of genetic erosion are widely known and international campaigning on plant genetic resources issues is well established. But most development agencies have yet to incorporate these concerns into their practical support for seed systems at the local level.

There is a trend towards relying on *in situ* conservation, including on-farm conservation of landraces and farmers' varieties, to complement conservation in gene banks, but there are problems with this approach. First, there is a lack of technical knowledge on the efficiency of on-farm maintenance of landraces for the conservation of genetic resources. Second, the potential tensions and synergies between genetic resources conservation on the one hand, and agricultural development on the other, are not well worked out.

The motives for on-farm conservation include both concerns about biodiversity conservation *per se*, and about development. Farmers may benefit from the greater stability of yield and multiple outputs provided by their traditional farming systems based on local varieties, as well as from greater independence from external suppliers. These benefits have to be weighed against the potential benefits of using modern varieties. On-farm conservation should be promoted only where farmers benefit, in the short or long term, or in special cases such as pilot studies, or where there is an over-riding national or global interest in conservation and the farmers concerned are duly compensated. There is a danger of replacing one straitjacket (dependence for preservation on gene banks and seed companies) with another (museum-like preservation of old varieties). A range of different strategies should be pursued, each of which can contribute to enhanced plant genetic resources conservation. These include re-oriented plant breeding; and local seed production.

As we saw in Chapter 1, varieties developed by farmers within communities are different in a number of important respects to modern varieties bred by formal sector agricultural researchers. Among the differences are the degree of intra-varietal variation. The most appropriate degree of varietal distinctness and stability is influenced significantly by the kind of agro-ecosystem prevailing. If environmental conditions are relatively stable but complex, the need is for a large number of distinct varieties, to maximise productivity by slotting each into a given micro-zone. However, if growing conditions are highly variable (in terms of rainfall, etc.) rather than highly complex, a large degree of intra-varietal variation is more useful than a large number of distinct varieties, to increase the chances of part of the crop producing a harvest. Agencies have to ensure that in such variable conditions support for local seed systems does not displace intra-varietal variation with introduced finished varieties (or produce the same effect through the introduction of formal sector plant breeding methods).

Farmers in marginal areas often prefer local varieties to those imported from outside the area. This is because local varieties are frequently better adapted to the area, and so perform better: it is difficult for formal sector researchers on research stations that are often located in other agro-ecological zones to breed varieties that are suitable for this kind of environment. However, outside highly marginal environments, farmers are often unwilling to maintain landraces and farmers' varieties on-farm because they yield lower than other available material. Farmers' processes of in-field selection and breeding do not necessarily maintain the old landraces: like formal sector plant breeders, they too are looking for the best performance, and maintenance of the genetic base of their farming system is of secondary concern. The need here is for genetic improvement of these farmers' varieties and other ways of promoting their effectiveness (Worede in Cooper *et al.*, 1992).

Related to this, farmers will not hesitate to introduce exotic material themselves if it is available and more satisfactory that what is available locally. Clear evidence of this is provided in Linnemann and Siemonsma (1989), Berg *et al.* (1991), Mooney in Cooper *et al.* (1992) and from our own case study of SCF in The Gambia, where farmers spread the introduced rice variety *Peking* with alacrity. The agency involved became concerned at the implications for genetic diversity, not the farmers themselves.

Agencies must assess carefully the characteristics of any variety they seek to introduce to an area, as the available modern varieties of many crops are not relevant for farmers in marginal environments. The available evidence suggests that insufficient attention has been paid to this to date. In particular, many NGOs have been too willing to assume that farmers' varieties are the solution to all farmers' variety problems.

The value of different varieties to farmers depends not only on their physical characteristics but also on the prices they fetch in local markets. Therefore, how any introduced varieties, whether modern varieties or farmers' varieties, will fit into the local farm economy must be established before any decisions are made about which varieties to make available.

This is one example of the important influence the farm household economy has on farmers' attitudes to variety choice: elsewhere in this book, we have discussed the influence of the economic function of the crop (food, cash, non-grain use, etc.) on variety choice. The diverse uses households make of different crops and varieties is another reason why it is important to offer as many different varieties as possible for farmers to choose from. In many situations, introducing modern varieties into the farming system can be helpful, so long as they are seen as part of a range of plant genetic resources for farmers to employ for different purposes.

There are substantial differences in the attitudes and methods of different agencies in this area of genetic resources conservation. There is also much rhetoric from all the actors: seed companies; advocacy groups; NGOs, etc. However, it is by no means clear that all agencies supporting local seed systems are improving the genetic resources base of local farming systems.

It is important to remember that the characteristics of the modern varieties that are currently available are not inherent but the result of the approach to variety development pursued by formal sector plant breeders. This is especially true of many of their disadvantages for small farmers, such as dependence on external inputs. It is unrealistic to ignore formal sector plant breeding because the formal sector can use many useful techniques that farmers cannot. The need is rather to decentralise plant breeding and to re-orientate breeding objectives better to serve the needs of small farmers (see pp 115-118 below).

The introduction of farmers' varieties is often of little immediate economic benefit to individual farmers. This is often the result of the common practice of subsidising the price of seed and other inputs for modern varieties. At the national level, pressure needs to be applied to ensure that the effects of subsidies are neutral on the relative benefits of using farmers' varieties and modern varieties.

At the international level, there is a need for continued pressure to ensure that plant genetic resources are not monopolised under legislation on plant breeders' rights and intellectual property rights.

In conclusion, while genetic erosion is undoubtedly a threat to the long-term sustainability of global agriculture, it is not sufficient to press for *in situ* conservation of genetic diversity and it is incorrect to assume that it makes economic sense for farmers to maintain on-farm a static gene pool, based around local varieties.

Instead, the need is to make a wide range of material available for farmers to choose from, including unfinished varieties displaying intra-varietal variation. Any more targeted promotion of specific varieties must be based on a very careful assessment of local environmental conditions, the local farm household economy and the characteristics of the varieties.

The Role of Government and Donors

The paramount influence of government policy on the success of farmers' and NGOs' seed activities has been highlighted many times in this book. So we end with an 'Agenda For Action' for government policy-makers, and for the donors that support them and many of the NGOs working in developing countries today.

Government Support for Local Seed Systems

Governments should recognise the value and effectiveness of small farm communities controlling and operating their own seed systems, and they should provide as much support as possible for communities to enable them to do this. The critical policy areas where change could help are—seed legislation, seed pricing, co-ordinated seed policy, plant breeding, seed technology research, and institutional linkages:

Seed legislation. The high standards set by the International Seed Testing Association (ISTA) are often not relevant to small farmer seed users. Furthermore, they are not always met by formal sector seed institutions, due to lack of resources for seed testing and problems with maintaining quality after testing in subsequent handling and storage. Therefore, one important way in which governments could encourage local seed systems is to relax national seed quality standards, retaining emphasis only on those aspects that are of real relevance to small farmer seed users. In this way, communities could officially trade as 'seed' the material that they produce, enabling them to increase sales and to charge realistic prices, as well as enabling them to reduce production costs (through not having to observe all the in-field inspection and subsequent testing that is required for ISTA standards). As we saw in Chapter 2, this is unlikely to lead to a real reduction in the actual quality of seed offered.

Seed pricing. One of the main limitations on the long-term economic viability of many of the local seed systems being supported by NGOs is government intervention in agricultural price-setting, where this means official seed prices do not reflect the full costs of seed production. An important aspect of this is the tying of agricultural subsidies and credit programmes to the use of modern varieties: this can artificially promote the use of modern varieties and severely restrict the viability of local seed projects geared to supplying farmers' varieties.

Governments need to ensure as far as possible that official seed prices reflect seed production costs and that subsidy and credit programmes do not distort the relative balance between modern varieties and farmers' varieties in a way which will harm the long-run sustainability of local seed systems. Indeed, in certain circumstances the long-term national interest might best be served by subsidies which favour farmers' varieties.

Co-ordinated seed policy. A number of developing country governments have introduced seeds-related legislation and have intervened in seed market management. However, few have any means of ensuring that the many other areas of government policy with an influence on the seed sector take the needs of the sector into account. The development of the sector could therefore be facilitated by co-ordination mechanisms, such as national seeds boards or seeds policy units within the Ministry of Agriculture. This could promote the sector at policy level and minimise the conflicting signals given by different government departments to farmer seed growers and to seed users.

Such units could also facilitate the development of national plans for the conservation and sustainable use of biological diversity and the integration of these plans into agricultural development policy. Such action is called for in Agenda 21, agreed at the 1992 UN Conference on Environment and Development, and is a requirement for countries that ratified the UNCED Convention on Biological Diversity.

Plant breeding. 'Plant breeders adapt our crops to the needs of development and the needs of commercial or legal concerns. They therefore often hold the key to development choices' (Berg *et al.*, 1991:1). Formal sector plant breeding could be made more relevant to small farmers in marginal environments—at least to some extent—by re-orienting plant breeding methods to take account of farmers' needs. This is now widely accepted. Changes in the international agricultural research centres and national agricultural research systems that serve developing countries should come from within and via the influence of governments and donors (GRAIN in Cooper *et al.*, 1992). Although farmers' organisations and NGOs representing farmer interests are calling for a radical change in direction, farmers themselves usually do not have the power to secure such changes.

As Biggs (1981) points out, it is relatively simple to identify the ideal institutional model for formal sector agricultural research. However, external factors—such as the professional objectives of scientists, the interests of international donors and different client groups, and national development goals—have a critical influence on the actors within this model and thus on the relative strength of the key linkages. In many cases, this has meant that the ideal model has not operated in practice. The type of on-farm client-oriented research approach described by Kaimowitz (1990) is one way of reducing these imperfections but much more needs to be done to make agricultural research genuinely adaptive and participatory.

Farmers' capacity for plant breeding needs to be investigated properly: few objective investigations of this have been made so far (but see Chambers, Pacey and Thrupp, 1989 for examples). However, the available evidence suggests that farmers themselves have considerable capacity to develop new varieties. The need from the formal research system is therefore for advanced material for selection at farm level and not only for finished varieties. Formal sector research institutions should be encouraged to allocate resources and tools to strengthen community innovation in genetic conservation and breeding. Communities need support because, as pressure mounts on farmers to increase production, they need access to new techniques and a wider range of germplasm.

The formal research system needs to continue with basic plant breeding for the crops and environments of most relevance to small farmers, as this tends to be neglected by private sector institutions with the capacity to do this kind of work. Within basic plant breeding work, formal research institutions need to change their methods of working in at least three ways. They need to include informal approaches, which involve the participation of farmers, as well as the more conventional scientific approaches. They need to include farmer preferences (drought resistance, low external input requirement, taste, storage, pest and disease resistance, non-grain yield, etc.) as well as scientific concerns in the selection of attributes. And they need to breed for diversity within varieties in some cases and not only uniform, stable varieties. This approach has secondary benefits in terms of reducing the time and cost of formal sector plant breeding: it is usually the stabilisation of varieties, once they have been selected, that takes time and this is a feature that in most circumstances small farmers do not require. They prefer instead to have some element of variability in the varieties that they plant, to cope with the variability inherent in the more marginal environments in which most of them farm. This approach has been termed 'integrated plant breeding' and is being actively promoted by a number of institutions (see, for example, Berg *et al.*, 1991).

Seed technology research. There is a need for more research on appropriate local level techniques and technologies for seed care. Specific areas in which further research and development work is needed were outlined above (pp. 28-30). As well as seed production practices, work is particularly needed on appropriate seed storage techniques and technologies. This has tended to be neglected to date, in favour of large-scale, centralised storage systems.

Institutional linkages. There are some seed sector functions that governments must continue to perform; the need is for these to be organised in a way that is more accessible to small farmers.

Basic plant breeding work oriented to the needs of small farmers in marginal and variable environments is definitely needed, as described above. The need for other functions depends on the context. If there is good infrastructure and the varieties bred by the formal sector are relevant, an initial supply of finished varieties to feed into local seed diffusion systems will speed up the process of diffusion. But there is no point in doing this if the supporting infrastructure is not there and the varieties are not relevant. Similarly, if the local seed system is already developing along relatively formal lines, governments can hasten the development of commercial seed production by providing decentralised quality control and certification services—but these will not be relevant if systems are localised and not monetised.

Whether governments seek to serve small farmers directly or to service NGOs as frontline agencies, depends in part on the historical development of services in small farm areas. In certain areas governments have handed over most seed service responsibilities to NGOs (for example, in parts of Ethiopia, and in The Gambia and Nepal). Whatever their role, governments should encourage conditions favourable to the development of community seed systems.

Facilitating the Work of NGOs

Governments should recognise the useful role NGOs can play in local seed systems, and facilitate this as far as possible. Areas where changes could help are national seed plans and institutional linkages:

National seed plans. Governments could include NGO seed activities in national seed sector planning and policies. The limited scale on which this has been done so far reflects the fact that relationships between governments and NGOs in the wider development process are still evolving. So far, governments have tended to share discrete tasks of implementation with NGOs, but they have not encouraged NGOs to contribute to policy-making and other forms of innovation; this could bring significant benefits (Bebbington and Farrington,

1993).

Institutional linkages.   Governments could encourage formal sector agricultural research and extension institutions to build linkages with NGOs working with small farmers. Such links could provide both valuable feed-back, based on NGOs' direct contact with farmers, and benefits to NGOs from the developments in plant breeding and other agricultural research that come out of these institutions. The need for this is increasingly recognised (see, for example, Farrington *et al.*, 1993) now that it is generally accepted that agricultural innovation requires inputs from many different institutions, including from farming communities themselves. More successful innovation thus requires the links in the 'technology triangle' of farmers-researchers-extensionists to be strengthened (Kaimowitz, 1990).

So far links have been almost universally problematic. Sometimes this is because of antagonism of government agencies towards NGOs, sometimes because of the public sector's chronic under-capacity and consequent inability to perform. Many government agencies have been primarily interested in forming links with NGOs for service delivery; but NGOs are capable of much more than this. Government organisations need to be open and flexible to capitalise on what NGOs have to offer.

Problems have also arisen because of antagonism on the part of NGOs. There is a fundamental dilemma here for those NGOs that see themselves as being alternatives to government. The appropriateness of this self-defined role depends on difficult questions about, for example, the merit of development paths based on different ideologies, and cannot easily be summarised. It is worth pointing out, however, that there are many cases where NGOs' isolation from government is not based on ideological grounds and is simply the result of poor planning and communication. This can result in considerable wasted resources and duplication of effort, as we saw in the case of emergency seed supply in Mozambique. In such situations, the efforts of some developing country governments to bring NGOs into mainstream development appear justified.

Changes in the attitude of governments to local seed systems and the agencies that support them will become increasingly important as economic reform emphasises market mechanisms and the role of the private sector (Smith and Thomson, 1991). The market will not sustain local seed systems in the more remote and marginal small farm areas: transactions costs, especially in the collection of information about different varieties, are too great for formal markets to supply the seeds needed. Hence, non-market organisations are indicated. And—given the variability of local needs for different kinds of seeds—small-scale and local-level enterprises, attuned to local conditions, are likely to have advantages over larger and more centralised operations.

In the era of cuts in the State sector, it is also important to work out the most effective role for what is left of government seed services after budget cuts and privatisation. The private sector is rarely interested in the seed needs of small farmers and budget cuts in many countries mean that, although a government seed service still exists, it does not have the resources and motivation to perform effectively. The NGOs in The Gambia provide good examples of how government seed services can link in with local agencies. Again, the most appropriate strategy depends on what already exists within local communities and on what roles NGOs are already performing.

The Role of Donors

Donors have an important role to play in enabling governments to take the issues discussed above into consideration when planning seed policies and programmes. Their role is not limited to imposing conditions attached to aid programmes. A critical role for donors is in funding networking and information exchange activities that will allow NGOs (and possibly even communities themselves) to learn from each others' experiences and to have access to the best available information when designing and implementing their own seed projects. This kind of activity is becoming increasingly significant all over the developing world, and we have seen in this book the importance of GRAIN's and RAFI's roles in promoting South-based seeds activities.

There are also two critical areas to which donors must consider contributing in the international arena. The first is supporting developing country governments in their lobbying against the introduction of intellectual property rights for genetic resources. Plant breeders rights and patents prevent on-farm trials of new and unfinished varieties being carried out by agencies other than the owner of the genetic material. Were they to be instituted widely, material would increasingly fall into the hands of private commercial seed companies and small farmers would be denied the opportunity to test and to modify material to their own specific conditions. This would prevent most local seed systems using improved material from outside sources. It would also prevent the potentially fruitful combination of local and high technology, precluding both locally adapted modern varieties (through restrictions on the availability of germplasm) and enhanced farmers' varieties (by restrictions on the availability of new plant technologies). These are the two most promising areas for sustainable improvements in small-farm agriculture through the use of plant genetic resources. Lobbying against this development requires co-ordinated international effort, of the kind donors can provide, as individual governments face a dilemma: without being certain that most other governments will refuse to permit patents and royalties for seed, the best option for protecting their own country's genetic resources and plant breeding research work is for governments to implement plant breeders rights and patents themselves.

The second critical area, for which donors are directly responsible via their contributions to the Consultative Group on International Agricultural Research, is the re-orientation of the international agricultural research centres' plant breeding work better to meet the seed needs of small farmers in developing countries (see above, p. 116). This is another issue over which individual developing country governments have little influence and which requires, instead, concerted international effort.

Support for local seed systems is not a complete substitute for conventional seed programmes in developing countries. Rather it is a complement: other systems are still relevant for some markets and crops. However, local seed systems are important and this is increasingly recognised by governments and donors. In the immediate future, a new *people-centred seed strategy* is needed to meet the seed needs of the majority of small farmers in developing countries who are outside the high potential agricultural zones.

The long-term aim must be to create conditions in which the different constituent parts of national seed supply systems can identify their own roles, interact with other components, and modify both roles and interactions in a dynamic fashion as conditions change. Over time, this should have a self-reinforcing effect: as on-farm evaluation of breeding material increases, due to increased local seed production, so varieties more suited to small farmers' needs will be developed. Thus the overall benefits of supporting multi-institutional approaches to seed supply will be far greater than simply a physical increase in the amount of seed within local communities. It is time for all agencies involved in supporting the efforts of small, poor farmers in the developing world to recognise the enormous potential of local seed systems to improve the sustainability of agriculture in the long run.

Appendix 1
Some Important Aspects of Seed Technology

A 'variety' of seed can be likened to a particular 'brand' of a product: in the case of seed, it is a brand which has been bred to have individual genetic characteristics, different to those of other brands of the same species. To be released via the formal research system, a variety must be Distinct, Uniform and Stable (the DUS conditions), and it must show value for cultivation and use. However, farmers' varieties can be relatively unstable as it can take many years and relatively sophisticated breeding techniques to stabilise varieties of some crops. Therefore, farmers' varieties can be relatively non-uniform, with considerable variation within as well as between varieties.

Different crops have different breeding systems and this determines the ease of maintaining the genetic integrity of a variety (see Appendix Table 1.1). The majority of cereal crops, including rice, wheat and barley, are normally self-pollinated, as are virtually all legumes (groundnuts, beans, soyabeans, etc.). The exceptions which are cross-pollinated include maize, sorghum, millet, sunflower and pigeon peas.

Self-pollinated crops are easy to maintain because they exist naturally as pure lines and any variability which occurs, for example from mutations or mechanical contamination, is visible and can be eliminated by roguing. They require isolation only to the extent of a physical barrier to avoid confusion with adjacent crops at sowing and harvest time.

Cross-pollinated crops are more difficult to manage because they are intrinsically variable as they are prone to contamination by foreign pollen, so seed crops have to be isolated either by space or time from others of the same species. If contamination does occur, it is less easily detected due to the variability which already exists within the variety, which tends to increase with successive multiplications of the crop.

The traditional varieties of cross-pollinated crops are open-pollinated populations, i.e. pollination is not controlled, but the formal sector attempts to restrict variability within cross-pollinated crops by breeding composite or synthetic varieties. The ultimate solution is to produce F1 hybrids by controlled crossing of parent lines, which is a labour and management intensive activity. Hybrid seed has to be bought fresh every year but it can produce higher yield by capitalising on hybrid vigour.

There are two aspects to seed quality, both of which are required for seed to contribute fully to crop yield. The first is its genetic potential (the genetic information contained within the seed itself). This is controlled by inspection in the field of the growing seed crop and removal of off-types by roguing. The second aspect is its physiological quality. This is controlled by sampling seed

Appendix Table 1.1: Important biological features of major crop species

	Hybrid Maize	Open Pollinated Maize	Sorghum/Millet	Wheat	Rice	Beans	Groundnuts
Breeding system	Controlled pollination	Cross pollination	Intermediate	Self pollination	Self pollination	Self pollination	Self pollination
Sowing rate per ha	Medium (20 kg)	Medium (20 kg)	Low (10 kg)	High (100 kg)	High (50 kg)	High (100 kg)	High (125 kg)
Multiplication factor	High (100)	High (100)	High (100)	Low (25)	Medium (50)	Medium (50)	Very low (<10)
Rate of deterioration	Very rapid	Rapid	Medium	Slow	Slow	Very slow	Very slow
Frequency of purchase	Annual	2 years	3 years	4 years	4 years	Variable	Variable
Availability of modern varieties	Many	Many	Few	Many	Many	Few	Few
Justification for purchase	Essential	Good	Variable	Poor	Poor	Poor	Very poor

for its germination capacity, purity, health and moisture content. In formal systems, both these aspects of seed quality are assured by seed certification. Certification requires that different generations of seed are identified and that the total number is limited. The standard nomenclature for this is:

- Breeder seed : limited quantities produced by plant breeders;
- Foundation seed (or basic seed or source seed) : produced under careful supervision from breeder seed;
- Registered seed (or Certified 1) : produced on a large scale by seed growers for sale for crop production
- Certified seed (or Certified 2) : the only subsequent multiplication that is recognised as seed rather than grain.

Source: Cromwell, E., E. Friis-Hansen and M. Turner (1992).

Appendix 2
Seed Needs Identification Survey —
Key Questions

The aim of the initial needs identification survey is to obtain all the information necessary to plan the elements of support for local seed systems listed in Box 8.1.

Information Sources and Data Gathering Method
As a starting point, a *review of the secondary data* held by agencies such as the Department of Agricultural Research and Department of Agriculture will yield much useful information. The basic information required at this stage will be:

- list of officially recommended crop varieties;
- results of official varietal trials and demonstrations;
- official seed demand and supply estimates (for relevant region);
- official seed prices (for relevant region).

There may also be a considerable amount of information available from official surveys, etc. relating to the more detailed information needs listed below.

However, the most important source of information is first-hand contact with communities, and the farmers within them, in the proposed project area.

Semi-structured group interviews are useful in gaining an overall view of the seed problems and related issues in the area and in sensitising communities to the intentions of the agency. The number carried out will depend on the size of the proposed project area; at least five will be necessary, to allow trends and differences to emerge. The type of questions to be asked are listed below. However, it is important that they are asked in an open question format, to allow the communities to express all the concerns they have regarding seed supply, some of which may otherwise be missed.

These should be followed up with *semi-structured key informant interviews*, both with farmers and with staff of relevant local institutions. For the farmer interviews, the number will again depend on the size of the area included in the project; it should be at least sufficient to capture any significant variations within the communities where the semi-structured group interviews have been conducted (10 per group interview could be an approximate guide). It is essential that members of vulnerable groups within the community are included in these interviews. Ideally, the interviews should be conducted just before planting time, when seed issues are uppermost in farmers' minds, but before the peak labouring season when time for answering questions will be short.

The interviews with local institutions should cover all those with an influence on the local seed system including: local Ministry of Agriculture offices; input supply offices; other agriculture and community development projects; private traders/local market vendors and traditional authorities within the community.

Information Needed

I Agro-ecosystem

* Rainfall, amount and variability;
* Local cropping pattern, including varieties used;
* Seasonal calendar of crop and variety planting, cultural management and harvesting;
* Crop and variety yields and factors influencing them;
* Traditional seed care practices: seed selection, seed treatment, seed storage;
* Seasonal calendar of field and store disease and pest occurrence by crop and variety.

II Farm household economy

* Economic function of different crops within the farming system (food, other domestic use, cash, etc.);
* Sufficiency of domestically-produced crops for household food and seed needs;
* Main felt needs for better standard of living and increased agricultural production;
* Seed sources, including use made of MVs compared to FVs and qualitative assessment of the various sources of seed;
* Returns to household resources, especially labour, in off-farm activities.

III Farmers' seed needs

* Varieties of seed required;
* Quality of seed desired;
* Quantity of seed required;
* Time of year in which seed required;
* Preferred source of seed;
* Price prepared to pay for seed.

For each of these questions, information should be obtained about what farmers' want compared to what is currently supplied, and about distinctions in this between households. Detailed questions that can be asked for all of the above are described Chapter 6 of Cromwell, Friis-Hansen and Turner (1992).

IV *Organisational opportunities*
- Current seed sources;
- Existing community self-help structures, traditional or introduced;
- Existing links with outside agencies, including agricultural research and extension services, input supply agencies, marketing authorities, other development agencies;
- Farmers' suggestions for the organisation of the project.

The aim is to make an accurate assessment of how support for local seed systems can be organised in a way that improves access to seed while building on existing community strengths.

V *Supplementary questions if seed bank to be established*
Who will be responsible for:
- keeping the money and accounts of the bank;
- keeping the store records and records of what seed is needed by the bank;
- buying and selling the seed;
- looking after the seeds in store;

What equipment will need to be bought for the bank? (item, quantity, estimated price, source);

Where will the seed be stored? In what? Will it be treated?;

How much money can the community raise for the bank? Who will the money be raised from? Who will be responsible for raising it?

Sources: Cromwell, Gurung and Urben, 1992; Cromwell, 1992; Cromwell, Friis-Hansen and Turner, 1992; Cromwell and Zambezi, 1992; Sperling *et al.*, 1992; Singh, 1990; Renton, 1990; Velasquez and Lewerez in CIAT, 1982.

Appendix 3
Sources of Technical and Organisational Advice on Local Seed Supply

The list that follows is a guide to organisations with experience in the planning and operation of local seed projects or in related policy issues, based on material used by ODI in carrying out the research on which this book is based. As such, it is not exhaustive.

International Centre for Tropical Agriculture (CIAT)
Seeds Unit
Apartado Aereo 6713
Cali
Colombia

Food and Agriculture Organisation of the United Nations (FAO)
Seeds and Plant Genetic Resources Service
via delle Terme di Caracalla
00100 Rome
Italy

Genetic Resources Action International (GRAIN)
Jonqueras 16-6-D
E–08003 Barcelona
Spain

Development Research Centre (IVO)
PO Box 90153
5000 LE
Tilburg
The Netherlands

Farmers–Scientists Partnership for Agricultural Development (MASIPAG)
c/o MASIPAG Secretariat (Nitz D. Abergas) 12
11th Avenue
Murphy
Cubao
Quezon City
Philippines

Natural Resources Institute (NRI)
Central Avenue
Chatham Maritime
Kent
ME4 4TB
UK

National Institute for Agricultural Botany (NIAB)
Huntingdon Road
Cambridge
CB3 OLE
UK

Rural Advancement Foundation International (RAFI)
Box 188
Brandon
Manitoba R7A 5Y8
Canada

Seed Technology Unit
School of Agriculture
University of Edinburgh
West Mains Road
Edinburgh EH9 3JG
UK

Winrock International Institute for Agricultural Development
1611 N Kent Street #600
ARLINGTON
Virginia 22209
USA

References

Action Aid–Nepal (1991), *Annual progress report July 1990–June 1991.* Kathmandu: Action Aid–Nepal.

ARD (1991), 'Décentralisation, gouvernance et gestion des ressources naturelles renouvables: options locales dans la république du Mali'. Paris: Associates in Rural Development Inc for Club du Sahel.

Bal, S.E. and J.E. Douglas (1992), 'Designing Successful Farmer-Managed Seed Systems', *Development Studies Paper Series.* Arkansas: Winrock International Institute for Agricultural Development.

Bebbington, A. (1989), 'Institutional options and multiple sources of agricultural innovation: evidence from an Ecuadorean case study', *Agricultural Administration (Research and Extension) Network Paper* No.11. London: Overseas Development Institute.

Bebbington, A. (1990), 'Sustainable resource use lost and sought: farm, region, and peasant organization in the central Ecuadorean Andes', Paper presented to the annual conference of the Society for Latin American Studies, Oxford, March–April.

Bebbington, A. and J. Farrington (1993), 'Governments, NGOs and Agricultural Development: Perspectives on Changing Inter-Organizational Relationships', *Journal of Development Studies* (forthcoming).

Berg, T. (1992), 'Indigenous Knowledge and Plant Breeding in Tigray, Ethiopia', *Forum For Development Studies*, No.1.

Berg, T. A. Bjørnstad, C. Fowler and T. Skrøppa (1991), 'Technology options and the gene struggle', NORAGRIC *Occasional Paper.* Ås, Norway: Norwegian Centre for International Agricultural Development, Agricultural University of Norway.

Biggs, S. (1981), 'Institutions and decision-making in agricultural research', *Agricultural Administration (Research and Extension) Network Discussion Paper* No.5. London: Overseas Development Institute.

Borton, J., N. Nicholds, J. Shoham, and M.O. Mukhier (1992), *An evaluation of Concern's 1990–91 emergency programme in Kosti Province, Sudan.* London: Overseas Development Institute.

Brown, L.D. (1991), 'NGOs as Bridging Organizations: a Complex Role', *IMPACT* No.14.

Buckland, J. and P. Graham (1990), 'The Mennonite Central Committee's experience in agricultural research and extension in Bangladesh 1973–1990', *Agricultural Administration (Research and Extension) Network Paper* No.17. London: Overseas Development Institute.

CESA (1991), 'Un aporte a la autogestión campesina: autoabastecimiento de semillas de calidad'. Quito: CESA.

Chambers, R., A. Pacey and L.A. Thrupp (eds.) (1989), *Farmer first: farmer* *innovation and agricultural research*. London: Intermediate Technology Publications.

CIAT (1982), *Proceedings of the Conference on Improved Seed for the Small Farmer 9–13 August, Cali, Columbia*. Cali, Colombia: CIAT.

CIAT (1991), *Annual Report*. Cali, Colombia: CIAT.

CIAT (1991), *Setting a Seed Industry in Motion: a non-conventional, successful approach in a developing country (Bolivia)*, CIAT Working Document No.57. Cali, Columbia: CIAT.

CIAT (1992), *Farmer participatory research and the development of an improved bean seed strategy in Rwanda*. Paper prepared for workshop on Farmer Participatory Research, 17–19 February, Addis Ababa, Ethiopia.

Concern (1991), *Final report on the emergency seed programme 1991–South White Nile province, Sudan*. Kosti and Dublin: Concern.

Cooper, D., R. Vellvé and H. Hobbelink (1992), *Growing diversity: genetic resources and local food security*. London: Intermediate Technology Publications.

Corbett, J. (pers. comm.), Interview with Mr Justin Corbett at ODI, London on 16 October 1992.

Cromwell, E., E. Friis-Hansen and M. Turner, (1992), 'The Seed Sector in Developing Countries: a framework for performance analysis', *Working Paper No.65*. London: Overseas Development Institute.

Cromwell, E.A. (ed.) (1990), 'Seed diffusion mechanisms in small farmer communities: lessons from Asia, Africa and Latin America', *Agricultural Administration (Research and Extension) Network Paper No.21*. London: Overseas Development Institute.

Cromwell, E.A. and B. Zambezi (1992), 'The performance of the seed sector in Malawi: an analysis of the influence of organizational structure, *Research Report*. London: Overseas Development Institute.

Cromwell, E.A., S. Gurung and R. Urben (1992), *Koshi Hills Agriculture Project Seeds Programme 1987–91: Impact Assessment*, Dhankuta: Koshi Hills Development Programme.

DANAGRO (1988), *Mozambique* Report for SADCC Regional Seed Production and Supply Project. Copenhagen: DANAGRO.

Dickie, A. (pers. comm.), Letter from Alex Dickie dated 9 September 1992.

Dulude, M. (1991), *Progam Update and Ongoing Proposal for the Seeds of Survival Programme in Ethiopia for Fiscal Year 1991/92*. Ottawa: Unitarian Service Committee.

Ellis, F. (1988), *Peasant Economics*, Cambridge: Cambridge University Press.

Farrington J., A. Bebbington, D. Lewis and K. Wellard (1993), *Reluctant partners? NGOs and the State in sustainable agricultural development*. London: Routledge.

Farrington, J. and S. Biggs (1990), 'NGOs, Agricultural Technology and the Rural Poor', *Food Policy* Vol.16, No.1.

Filleton, P. (1989), 'Analyse des greniers semenciers mis en place à Timbuktu'.

Timbuktu: ACORD.

Fowler, A. (1991), 'The role of NGOs in changing state-society relationships: perspectives from Eastern and Southern Africa', *Development Policy Review* Vol.9, No.1. London: Overseas Development Institute.

Gaifami, A. (1991) *Crocevia's seed project in Niassa, Mozambique*. Rome: Centro Internazionale Crocevia.

Gaifami, A. (1991a) *Gabinete de producao de sementes do niassa*. Rome: Centro Internazionale Crocevia.

Gilbert, E. (1990), 'Non-governmental organisations and agricultural research: the experience of The Gambia', *Agricultural Administration (Research and Extension) Network Paper* No.12. London: Overseas Development Institute.

Green, T. (1987), 'Farmer-to-farmer seed exchange in the Eastern Hills of Nepal: the case of Pokhreli Masino rice', *Working Paper* 05/87. Dhankuta, Nepal: Pakhribas Agricultural Centre.

Groosman T. (1991), 'Seed Industry Development: developing countries' experiences in different crops'. *Research Report* No.34. Tilburg, Netherlands: IVO.

Heisey, P. (ed.) (1990), 'Accelerating the transfer of wheat breeding gains to farmers: a study of the dynamics of varietal replacement in Pakistan', *CIMMYT Research Report* No.1. Mexico: CIMMYT.

Henderson, P. and Singh, R (1990), 'NGO–government links in seed production: case studies from The Gambia and Ethiopia', *Agricultural Administration (Research and Extension) Network Paper* No.14. London: Overseas Development Institute.

Herthelius, L-P., K. Pehrsson, H. Svensk and M. Greeley (1989) *Appraisal of request for Swedish assistance to the development of the Mozambican seed industry 1990–93*, Report of a consultancy mission for SIDA. Stockholm: Swedish International Development Authority.

Hossain, M. (1990), 'Bangladesh: economic performance and prospects', ODI *Briefing Paper*. London: Overseas Development Institute.

ICDA (1989), *ICDA Seeds Campaign*. Barcelona: ICDA Seeds Campaign.

Kaimowitz, D. (ed.) (1990), *Making The Link: Agricultural research and Technology Transfer in Developing Countries*. Boulder, CO: Westview/International Service for National Agricultural Research.

Kohl, B. (1991) 'Protected Horticultural Systems in the Bolivian Andes: a case study of NGOs and inappropriate technology', *Agricultural Administration (Research and Extension) Network Paper* No.29. London: Overseas Development Institute.

Linnemann, A. and J. Siemonsma (1989), 'Variety choice and seed supply by smallholders', *ILEIA Newsletter*, December.

Linnemann, A.R. and G.H. de Bruyn (1987), 'Traditional seed supply for food crops', *ILEIA Newsletter* Vol.3, No.2.

Low, A. (1986), *Agricultural development in Southern Africa: farm-household economics and the food crisis*. London: James Currey.

Marty, A. (1985), 'Crise rural en milieu Nord Sahélien et recherche coopérative', Doctoral thesis (available from ACORD, London).

MCC (1989): *Agriculture Programme Report*. Bangladesh: Mennonite Central Committee.

Merrill-Sands, D. and D. Kaimowitz (1990), *The technology triangle. Summary report of international workshop held in November 1989*. The Hague: ISNAR.

Michigan State University (1987), *Malawi Collaborative Research Support Project Detailed Annual Report*. East Lansing: Bean/Cowpea CRSP Management Office, Michigan State University.

Miclat-Teves, A.G. (1991a), *Overview of the seed sector in The Philippines*. London: Overseas Development Institute.

Miclat-Teves, A.G. (1991b), *Towards achieving sustainable agriculture: the seed programme in Occidental Mindoro*. London: Overseas Development Institute.

National Seed Board (1990), *Proceedings of Second National Seed Seminar 20–22 March, Kathmandu*. Kathmandu: National Seed Board, Ministry of Agriculture.

NEF (1988), 'Douenza Seed Bank Workshop, 19–20 January'. Douenza, Mali: Near East Foundation.

Nieuwkerk, M. (1987), *Evaluation of ACORD Mali Programme*, London: ACORD.

Nieuwkerk, M., D. Sylla and B. Thebaud (1983), *Mali Evaluation 1983*. London: ACORD.

Notes on MCC's seed activities from **Peter Graham**, Agriculture Administrator, MCC, Bangladesh.

ODI (1988), 'NGOs in development', ODI *Briefing Paper*. London: Overseas Development Institute.

Osborn, T. (1990), 'Multi-institutional approaches to participatory technology development: a case study from Senegal', *Agricultural Administration (Research and Extension) Network Paper* No.13. London: Overseas Development Institute.

PAC (1986), *Guidelines for the selection of farmer seed growers*. Dhankuta, Nepal: Pakhribas Agricultural Centre, Seed Technology Section.

Poey, F. and M. Perez (1991), *Philippines: the development of the rice and corn seed industry*. Manila: United States Agency for International Development.

RAFI (1986), *The Community Seed Bank Kit*. Pittsboro, North Carolina: Rural Advancement Foundation International.

Rajbhandary, K.L., D.N. Ojha and S.S. Bal (1987) *Hill seed supply through private producer-sellers*. Kathmandu: Seed Technology and Improvement Programme, Agricultural Research and Production Project, Ministry of Agriculture.

Renton, C.L. (1988), 'The role of non-governmental organisations in village-level seed supplies', MSc dissertation, University of Edinburgh.

Renton, C.L. (1988), *The role of non-governmental organisations in village-level seed supplies*. MSc dissertation in Seed Technology, University of Edinburgh.

Renton, C.L. (1990), *A village seed bank programme in eastern Sudan*. London:

ACORD.

Renton, C.L. (pers. comm.), Letter from Christian Renton dated 18 June 1992.

SEAN (1991), *Involvement of Private Sector in Seed Business: Open House Discussion 25–26 June, Kathmandu.* Kathmandu: Seed Entrepreneurs Association of Nepal/Agricultural Inputs Corporation/Seed Production and Marketing Project/GTZ.

Sevilla, E.P. (1987), 'Philippines' in Asian Productivity Organisation *Cereal Seed Industry in Asia and the Pacific,* Tokyo: Asian Productivity Organisation.

Singh, R. (1990), 'A preparatory note for rapid rural seed appraisal', Band Aid/Action Aid/University of Edinburgh Seed Technology Unit SEED Project, Addis Ababa, Ethiopia.

Smith, L. and A. Thomson (1991), 'Achieving a reasonable balance between the public private sectors in agriculture'. Paper presented at 21st international conference of Agricultural Economist, Tokyo, Japan, 22–29 August.

Sperling, L. and M. Loevinsohn (1992), 'The dynamics of adoption: distribution and mortality of bean varieties among small farmers in Rwanda', *Agricultural Systems* (forthcoming).

Sperling, L., U. Scheidegger, B. Ntambouura, T. Musungayi and J.M. Murhandikire (1992), *Analysis of bean seed channels in South Kivu, Zaire and Butare and Gikongoro Prefectures, Rwanda.* Rwanda: CIAT and Programme National Légumineuses.

Turner, M.R. (1990), *The Philippines seed industry.* Notes on a Lecture to MSc Seed Technology course, University of Edinburgh (mimeo).

USC (1988), *African Seeds of Survival Programme: project proposal.* Ottawa: Unitarian Service Committee.

USC (1990), *Report on Seeds of Survival Ethiopia.* Ottawa: Unitarian Service Committee.

Villegas, G. (pers. comm.), Interview with Dr George Villegas at ODI, London on 17 August 1992.

Wiggins, S. (1992), *Non-governmental organisations and seed supply in The Gambia.* Reading: Department of Agricultural Economics and Management, University of Reading.

Williams, T. and T. Osman Hadra (1992), *Kebkabiya seeds restocking project 1991: evaluation report.* Oxford, UK: OXFAM.

World Bank (1990), *World Tables 1989–90 Edition.* Washington DC: World Bank.

World Bank (1992), *World Development Report.* Oxford: Oxford University Press.

Guide to Further Reading

The lists that follow are intended as a guide to useful material available on sustainable local seed supply and related issues. The references listed have been selected from the material used by ODI in carrying out the research on which this book is based and, as such, are not exhaustive.

Sustainable Agriculture

Carr, S. (1989), 'Technology for Small-Scale Farmers in Sub-Saharan Africa', *Technical Paper* No.109. Washington: World Bank.

Kesseba, A. (ed.) (1989), *Technology Systems for Small Farmers: Issues and Options*. Boulder, Colorado: Westview.

FAO/Ministry of Agriculture, Nature Management and Fisheries of The Netherlands (1991), *The Den Bosch declaration and agenda for action of sustainable agriculture and rural development*. Report of the Conference, s-Hertogenbosch, The Netherlands, 15–19 April.

IAD (1991), 'Sustaining Agriculture With Few Inputs', *International Agricultural Development* May/June.

Shaikh, A., E. Arnould, K. Christophersen, R. Hagen, J. Tabor and P. Warshall (1988), *Opportunities For Sustained Development: Successful Natural Resource Management in The Sahel*. Washington: Energy Development International/USAID.

Institutional Issues In Agricultural Development

Ashford, T. and S. Biggs (1992), 'The Dynamics of Rural and Agricultural Mechanizations: The Role of Different Actors in Technical and Institutional Change', *Journal of International Development* Vol.4, No.4.

Bebbington, A. (1989), 'Institutional options and multiple sources of agricultural innovation: evidence from an Ecuadorean case study', *Agricultural (Research and Extension) Network Paper* No.11. London: Overseas Development Institute.

Biggs, S. (1981), 'Institutions and decision-making in agricultural research', *Agricultural Administration (Research and Extension) Network Discussion Paper* No.5, (further references included in bibliography). London: Overseas Development Institute

Biggs, S. (1989), 'A Multiple Source of Innovation Model of Agricultural Research and Extension', *Agricultural Administration (Research and Extension) Network Paper* No.6. London: Overseas Development Institute.

Brinkerhoff, D. and A. Goldsmith (1992), 'Promoting the Sustainability of Development Institutions: A Framework for Strategy', *World Development* Vol.20, No.3.

Chambers, R., A. Pacey and L.A. Thrupp (eds.) (1989), *Farmer first: farmer innovation and agricultural research*. London: Intermediate Technology Publications.

Hildebrand, P. (1988), 'Technology Diffusion in Farming Systems Research and Extension', *Hortscience* Vol.23, No.3.

Kaimowitz, D. (ed.) (1990), *Making The Link: Agricultural research and Technology Transfer in Developing Countries*, (associated Theme Papers report results of components of this ISNAR research project). Boulder, CO: Westview/International Service for National Agricultural Research.

Low, A. and S. Waddington (1990), 'Maize adaptive research: achievements and prospects in Southern Africa', *Farming Systems Bulletin: Eastern and Southern Africa* No.6.

Pineiro, M. 'Generation and Transfer of Technology for Poor, Small Farmers', Chapter in Kesseba, A. (ed.) (1989), *Technology Systems for Small Farmers: Issues and Options*. Boulder, CO: Westview.

Smith, L. and A. Thomson (1991), 'Achieving a reasonable balance between the public private sectors in agriculture'. Paper presented at 21st international conference of Agricultural Economist, Tokyo, Japan, 22–29 August 1991.

NGOs' Role in Rural Development

ANEN (1988), *Workshop on Improving Indigenous African NGOs/African Governments/Aid Agencies Collaboration in African recovery and Development*, Report of the Proceedings. Nairobi: African NGOs Environment Network.

Bebbington, A. and J. Farrington (1993), 'Governments, NGOs and Agricultural Development: Perspectives on Changing Inter-Organizational Relationships', *Journal of Development Studies* (forthcoming).

Bratton, M. (1990), 'Non-Governmental Organizations in Africa: Can They Influence Public Policy?', *Development and Change* Vol.21.

Brown, D. (1990), 'Rhetoric or Reality? Assessing the Role of NGOs as Agencies of Grassroots Development', *Bulletin* No.28. Reading: Department of Agricultural Extension and Rural Development, University of Reading.

Brown, D. (1991), 'NGOs as Bridging Organizations: a complex role', *IMPACT* No.14.

Clark, J. (1991), *Democratising Development: The Role of Voluntary Organizations*. London: Earthscan.

Copestake, J. (1990), 'The Scope For Collaboration Between Government and PVOs in Agricultural Technology Development: The Case of Zambia', *Agricultural Administration (Research and Extension) Network Paper* No.20. London: Overseas Development Institute.

Farrington, J. and S. Biggs (1990), 'NGOs, Agricultural Technology and the Rural Poor', *Food Policy* Vol.16, No.1.

Farrington, J. (1991), 'NGOs and GOs Get Acquainted', *ILEIA Newsletter* No.4.

Farrington J., A. Bebbington, D. Lewis and K. Wellard (1993), *Reluctant* *partners? NGOs and the State in sustainable agricultural development*. London: Routledge.

Fowler, A. (1991), 'The Role of NGOs in Changing State-Society Relationships: Perspectives From Eastern and Southern Africa', *Development Policy Review* No.9.

Gordon Drabek, A. (ed.) (1987), 'Development Alternatives: The Challenge For NGOs', *World Development* Vol.15 (Special Supplement).

Gilbert, E. (1990), 'Non-governmental organisations and agricultural research: the experience of The Gambia', *Agricultural Administration (Research and Extension) Network Paper* No.12. London: Overseas Development Institute.

Kohl, B. (1991), 'Protected Horticultural Systems in the Bolivian Andes: a case study of NGOs and inappropriate technology', *Agricultural Research and Extension Network Paper* No.29. London: Overseas Development Institute.

Jamela, S. (1990), 'The Challenges Facing African NGOs: A Case-Study Approach', *Voices From Africa* No.2.

Johnson, W. and V. Johnson (1990), *West African Governments and Volunteer Development Organisations: Priorities For Partnership*. Lanham, MD: University Press of America.

Korten, D. (1991), 'The Role of Non-Governmental Organizations in Development: Changing Patterns and Perspectives', chapter in Paul, S. and A. Israel (eds.) (1991), *Non-Governmental Organizations and The World Bank: Co-operation For Development*. Washington: World Bank.

Livernash, R. (1992), 'The Growing Influence of NGOs in the Developing World', *Environment* Vol.34, No.5.

Narayana, E. (1992), 'Bureaucratization of Non-Governmental Organizations: An Analysis of Employees' Perceptions and Attitudes', *Public Administration and Development* Vol.12, No.2.

NOVIB/Netherlands Ministry of Foreign Affairs (1989), *Bigger NG(D)Os in Eastern and Southern Africa: An Analysis of Constraints and Effects of Growth*. Report of a joint mission by NOVIB and the Co-Financing Programming Section of the Netherlands Ministry of Foreign Affairs.

ODI (1988), 'NGOs in Development', *ODI Briefing Paper*. London: Overseas Development Institute (revised February 1989).

Osborn, T. (1990), 'Multi-Institutional Approaches to Participatory Technology Development: A Case Study From Senegal', *Agricultural Administration (Research and Extension) Network Paper* No.13. London: Overseas Development Institute.

Renton, C. (1988), 'The role of non-governmental organisations in village-level seed supplies', MSc dissertation, University of Edinburgh.

Wellard, K., J. Farrington and P. Davies (1990), 'The State, Voluntary Organizations and Agricultural Technology in Marginal Areas', *Agricultural Administration (Research and Extension) Network Paper* No.15. London: Overseas Development Institute.

Community Organisations

ARD (1991), 'Décentralisation, gouvernance et gestion des ressources naturelles renouvables: options locales dans la république du Mali'. Report by Associates in Rural Development Inc. for Club du Sahel, Paris, France.

Bebbington, A. (1990), 'Sustainable resource use lost and sought: farm, region, and peasant organization in the central Ecuadorean Andes'. Paper presented to the annual conference of the Society for Latin American Studies, Oxford, March–April 1990.

Curtis, D. (1991), *Beyond Government: Organizations For Common Benefit*. London: Macmillan.

Fraser-Taylor, D. and F. Mackenzie (eds.) (1991), *Development From Within: Survival In Rural Africa*. London: Routledge.

Harper, M. (1992), 'The Critical Factors in the Success of Co-operatives and Other Group Enterprises', *Small Enterprise Development* Vol.3, No.1.

Maddock, N. (1992), 'Local Institutions and the Management of Development Projects', *International Journal of Public Sector Management* Vol.5, No.2.

Mishra, D. and T. Shah (1992), 'Analysing Organizational Performance in Village Co-operatives', *Small Enterprise Development* Vol.3, No.1.

Salole, G. (1991), 'Not Seeing The Wood For The Trees: Searching For Indigenous Non-Governmental Organizations in the Forest of Voluntary Self-Help Associations', *Journal of Social Development in Africa* Vol.6, No.1.

Genetic Diversity

Altieri, M. (1987), 'The Significance of Diversity in the Maintenance of the Sustainability of Traditional Agro-Ecosystems', *ILEIA Newsletter* Vol.3, No.2.

Appropriate Technology (1992), 'Losing The Landrace', *Appropriate Technology* Vol.18, No.4 (special issue on biodiversity).

Berg, T., A. Bjørnstad, C. Fowler and T. Skrøppa (1991), 'Technology Options and the Gene Struggle', NORAGRIC *Occasional Paper*. Ås, Norway: Norwegian Centre for International Agricultural Development, Agricultural University of Norway.

BOSTID (1992), *Conserving biodiversity: a research agenda for development agencies*. Washington, DC: National Academic Press.

Bunting, A. (1990), 'The Pleasures of Diversity', *Biological Journal of the Linnean Society* Vol.39.

Cooper, D., R. Vellvé and H. Hobbelink (1992), *Growing diversity: genetic resources and local food security*. London: Intermediate Technology Publications.

GRAIN (1991), 'Genes for Sustainable Development', *Briefing on Biodiversity* No.2. Barcelona: GRAIN.

van Hintum, J., L. Frese and P. Perret (1992), *Crop Networks: Searching for New Concepts in Collaborative Genetic Resources Management*. Papers of the EUCARPIA/IBPGR symposium held at Wageningen, The Netherlands, 3–6 December 1990. Rome: International Board for Plant Genetic Resources.

ICDA Seeds Campaign (1989), *Patented? Patenting Life Forms in Europe.* Proceedings of a Conference held in Brussels, 7–8 February. Barcelona: ICDA Seeds Campaign.

ITDG/New Economics Foundation (1991), *The Genetraders: security or profit in food production?* Proceedings of a Conference held in London, 14–15 April. London: Intermediate Technology Development Group and New Economics Foundation.

Jiggins, J. (1990), *Crop Variety Mixtures in Marginal Environments*, Gatekeeper Series No.SA19. London: International Institute for the Environment and Development.

Kloppenburg, J. (1988), *First The Seed: The Political Economy of Plant Biotechnology 1492–2000.* Cambridge: Cambridge University Press.

McNeely, J., K. Miller, W. Reid, R. Mittermeier and T. Werner (1990), *Conserving the world's biodiversity.* Washington: World Bank, The World Resources Institute, The World Conservation Union, Conservation International and World Wildlife Fund.

Richards, P. (1987), 'Spreading Risks Across Slopes: Diversified Rice Production in Central Sierra Leone', *ILEIA Newsletter* Vol.3, No.2.

RONGEAD (1990), 'Of Minds and Markets: Intellectual Property Rights and the Poor', *GATT Briefing* No.2. Lyon, France: RONGEAD.

Sandoval, V. (1991), 'Accepting Uncertainty in Choosing Varieties', *ILEIA Newsletter* Vol.7, No.4.

Solow, A., S. Polasky and J. Broadus (1992), 'On the measurement of biological diversity', *Journal of Environmental Economics and Management* (forthcoming).

Worldwatch (1992), *Life support: conserving biological diversity.* Washington, DC: Worldwatch Institute.

Technical Aspects of Seed Production

FAO (1978), *Improved Seed Production*, (a technical manual). Rome: UN Food and Agriculture Organisation.

OFSP (1988), *Seed Multiplication Manual For Extension Workers in the Gambian Seed Industry.* Dakar: On-Farm Seed Project.

RAFI (1986), *The Community Seed Bank Kit.* Pittsboro, North Carolina: Rural Advancement Foundation International.

Simmonds, N. (1979), *Principles of Crop Improvement.* London: Longman.

Thomson, J. (1979), *An Introduction to Seed Technology.* London: Hill.

Wellving, A. (ed.) (1984), *Seed Production Handbook of Zambia.* Lusaka: Department of Agriculture.

Traditional Seed Systems

Appropriate Technology (1992), 'Losing The Landrace', *Appropriate Technology* Vol.18, No.4, (special issue on biodiversity).

Berg, T. (1992), 'Indigenous Knowledge and Plant Breeding in Tigray,

Ethiopia', *Forum For Development Studies*, No.1, (description of community-managed seed banks in Tigray).

CIAT (1992), *Farmer participatory research and the development of an improved bean seed strategy in Rwanda*. Paper prepared for workshop on farmer participatory research, 17–19 February, Addis Ababa.

Cromwell, E. (ed.) (1990), 'Seed diffusion mechanisms in small farmer communities: lessons from Asia, Africa and Latin America', *Agricultural Administration (Research and Extension) Network Paper* No.21. London: Overseas Development Institute.

Cromwell, E., S. Gurung and R. Urben (1992), *The Koshi Hills Agriculture Project: an impact assessment of the seed programme*. Dhankuta, Nepal: Koshi Hills Development Programme.

Cromwell, E. and B. Zambezi (1992), 'The Performance of the Seed Sector in Malawi: An Analysis of the Influence of Organizational Structure', *Research Report*. London: Overseas Development Institute.

Green, T. (1987), 'Farmer-to-farmer seed exchange in the Eastern Hills of Nepal: the case of Pokhreli Masino rice', *Working Paper* 05/87. Dhankuta, Nepal: Pakhribas Agricultural Centre.

Heisey, P. and J. Brennan (1991), 'An Analytical Model of Farmers' Demand For Replacement Seed', *American Journal of Agricultural Economics* Vol.73, No.4.

ILEIA Newsletter (1989), 'Local Varieties Are Our Source of Strength and Health', *ILEIA Newsletter* Vol.5, No.4, (special issue on local seed systems).

International Ag-Sieve (1990), 'Saving Seed In Senegal', *International Ag-Sieve* Vol.3, No.5.

Linnemann, A. and G. de Bruyn (1987), 'Traditional seed supply for food crops', *ILEIA Newsletter* Vol.3, No.2.

Sperling, L. and M. Loevinsohn (1992), 'The dynamics of adoption: distribution and mortality of bean varieties among small farmers in Rwanda', *Agricultural Systems*, forthcoming.

Sperling, L., U. Scheidegger, B. Ntambouura, T. Musungayi and J. Murhandikire (1992), *Analysis of bean seed channels in South Kivu, Zaire and Butare and Gikongoro Prefectures, Rwanda*. Rwanda: CIAT and Programme National Légumineuses.

Tetlay, K., P. Heisey, Z. Ahmed and A. Munir (1991), 'Farmers' Sources of Wheat Seed and Wheat Seed Management In Three Irrigated Regions of Pakistan', *Seed Science and Technology* Vol.19, No.1.

Seed Projects and Programmes

Asian Productivity Organisation (1987), *Cereal Seed Industry in Asia and the Pacific*. Tokyo: Asian Productivity Organisation.

Bal, S. and J. Douglas (1992), 'Designing Successful Farmer-Managed Seed Systems', *Development Studies Paper Series*. Morrilton, Arkansas: Winrock International Institute for Agricultural Development.

Camargo, C., C. Bragantini and A. Monares (1989), *Seed Production Systems For Small Farmers: A Non-Conventional Perspective*. Cali, Colombia: CIAT.

CIAT (1982), *Proceeding of the conference on Improved Seed For the Small Farmer. 9–13 August, Cali, Colombia*. Cali, Colombia: CIAT.

CIAT (1991), *Setting a Seed Industry in Motion: a non-conventional, successful approach in a developing country (Bolivia)*, CIAT Working Document No.57. Cali, Colombia: CIAT.

Concern (1991), *Final Report On The Emergency Seed Programme 1991 – South White Nile Province, Sudan*, Kosti and Dublin: Concern.

Cooper, D., R. Vellvé and H. Hobbelink (1992), *Growing diversity: genetic resources and local food security*. London: Intermediate Technology Publications.

Cromwell, E. (1992), 'The Impact of Economic Reform on the Performance of the Seed Sector in Eastern and Southern Africa', *OECD Development Centre Technical Paper* No.68. Paris: Organisation for Economic Cooperation and Development, Development Centre.

Cromwell, E., E. Friis-Hansen and M. Turner (1992), 'The Seed Sector in Developing Countries: a framework for performance analysis', *Working Paper* No.65. London: Overseas Development Institute.

DANAGRO (1988), 'Regional Seed Production and Supply Project'. Report for Southern Africa Development Co-ordination Committee by DANAGRO, Copenhagen, (synthesis volume plus separate reports for each SADCC country).

Delhove, G. (1992), *Seed Programmes and Projects in ACP Countries*. Wageningen: Technical Centre for Agricultural and Rural Cooperation (CTA).

Douglas, J. (1980), *Successful Seed Programmes: A Planning and Management Guide*. Boulder, CO: Westview Press.

FAO (1987), *FAO Seed Review 1984–85*. Rome: UN Food and Agriculture Organization.

Groosman T. (1991), 'Seed Industry Development: developing countries' experiences in different crops', *Research Report* No.34. Tilburg, Netherlands: IVO.

Groosman, T., A. Linnemann and H. Wierema (1991), *Seed Industry Development in a North–South Perspective*. Wageningen: PUDOC.

Hamidullah, H. (1990), *Seed and Fertilizer Distribution in Afghanaid Project Areas in 1989 and Expected Yields in 1990*. Peshawar, Pakistan: Afghanaid.

Henderson, P. and R. Singh (1990), 'NGO-government links in seed production: cases studies from The Gambia and Ethiopia', *Agricultural Administration (Research and Extension) Network Paper* No.14. London: Overseas Development Institute.

Kelly, A. (1989), *Seed Planning and Policy For Agricultural Production*. New York: Belhaven Press.

McMullen, N. (1987), *Seeds and World Agricultural Progress*. Washington: National Planning Association.

Neumaier, M., C. Yu and C. Freire (1990), 'Avaliacao socio-economica de producao comunitaria de sementes em Rio Azul-Parana (Brazil)', *Boletin Tecnico-Instituto Agronomico do Parana* No.29.

Rajbhandary, K.L., D.N. Ojha and S.S. Bal (1987), *Hill seed supply through private producer-sellers*. Kathmandu: Seed Technology and Improvement Programme, Agricultural Research and Production Project, Ministry of Agriculture.

Renton, C. (1988), 'The role of non-governmental organisations in village-level seed supplies', MSc dissertation, University of Edinburgh.

Sharrock, S. and M. Bhattarai (1986), 'The Initiation of a Village Level Seed Production and Supply Programme With a Co-operative Society', *Technical Report*. Dhankhuta, Nepal: Pakhribas Agricultural Centre.

Swedish Seed Association (1989), 'Seed Production For the Small-Scale Farming Sector in Africa', *Journal of the Swedish Seed Association* Vol.99, No.4 (Special Issue).

Seed Banks

Berg, T. (1992), 'Indigenous Knowledge and Plant Breeding in Tigray, Ethiopia', *Forum For Development Studies* No.1., (description of community-managed seed banks in Tigray)

Filleton, P. (1989), 'Analyse des greniers semenciers mis en place à Timbuktu'. Timbuktu: ACORD.

Goyder, H. (1987), 'Darfur Seed Banks'. Paper presented at OXFAM Arid Lands Workshop. Oxford: OXFAM.

Near East Foundation (1988), 'Douenza Seed Bank Workshop, 19–20 January'. Douenza, Mali: Near East Foundation.

Renton, C. (1990), *A Village Seed Bank Programme in Eastern Sudan*. London: ACORD/ODA.

Williams, T. and T. Osman Hadra (1992), *Kebkabiya Seeds Restocking Project 1991: Evaluation Report*. Oxford: OXFAM.